THE TRADE TECHNICIAN'S SOFT SKILLS MANUAL

THE TRADE TECHNICIAN'S SOFT SKILLS MANUAL

Steve Coscia

DELMAR
CENGAGE Learning™

Australia • Brazil • Japan • Korea • Mexico • Singapore • Spain • United Kingdom • United States

DELMAR
CENGAGE Learning™

The Trade Technician's Soft
Skills Manual
Steve Coscia

Vice President, Career
and Professional Editorial:
Dave Garza

Director of Learning Solutions:
Sandy Clark

Senior Acquisitions Editor:
James DeVoe

Managing Editor: Larry Main

Senior Product Manager:
John Fisher

Editorial Assistant: Cris Savino

Vice President, Career and
Professional Marketing:
Jennifer Baker

Marketing Director:
Deborah Yarnell

Marketing Manager: Katie Hall

Production Director:
Wendy Troeger

Production Manager:
Mark Bernard

Content Project Manager:
Mike Tubbert

Senior Art Director:
Casey Kirchmayer

For product information and
technology assistance, contact us at **Cengage Learning
Customer & Sales Support, 1-800-354-9706**

For permission to use material from this text or product,
submit all requests online at **www.cengage.com/permissions**.
Further permissions questions can be e-mailed to
permissionrequest@cengage.com.

Library of Congress Control Number: 2010941262

ISBN-13: 978-1-111-31381-4

ISBN-10: 1-111-31381-4

Delmar
5 Maxwell Drive
Clifton Park, NY 12065-2919
USA

Cengage Learning is a leading provider of customized learning solutions with office locations around the globe, including Singapore, the United Kingdom, Australia, Mexico, Brazil, and Japan. Locate your local office at: **international.cengage.com/region**

Cengage Learning products are represented in Canada by Nelson Education Ltd.

To learn more about Delmar, visit **www.cengage.com/delmar**

Purchase any of our products at your local college store or at our preferred online store **www.CengageBrain.com**

Notice to the Reader

Publisher does not warrant or guarantee any of the products described herein or perform any independent analysis in connection with any of the product information contained herein. Publisher does not assume, and expressly disclaims, any obligation to obtain and include information other than that provided to it by the manufacturer. The reader is expressly warned to consider and adopt all safety precautions that might be indicated by the activities described herein and to avoid all potential hazards. By following the instructions contained herein, the reader willingly assumes all risks in connection with such instructions. The publisher makes no representations or warranties of any kind, including but not limited to, the warranties of fitness for particular purpose or merchantability, nor are any such representations implied with respect to the material set forth herein, and the publisher takes no responsibility with respect to such material. The publisher shall not be liable for any special, consequential, or exemplary damages resulting, in whole or part, from the readers' use of, or reliance upon, this material.

Printed in the United States of America
1 2 3 4 5 XX 13 12 11

Table of Contents

Foreword

I first met Steve Coscia on a cold December day in Chicago when he was conducting a two-day customer service seminar for a leading commercial real estate developer. I found Steve's personality and expertise quite refreshing—he spoke with unmistakable authority. Meeting Steve was serendipitous too.

My role in working with both secondary and postsecondary schools on a national level is to provide their teachers with professional development, curriculum, and standardized student outcome assessment. This made me acutely aware of soft skills as a required key curriculum component and that a well-thought-out and structured soft skills module was in fact missing from most coursework.

I realized that I was sitting with a guy who had already written a customer service handbook and who was past president of a National Speakers Association chapter. Perhaps he would develop a full program, including instructor resources, and be able to speak about the benefits of soft skills training in the classroom. At my urging, Steve wrote a lesson plan, instructor's guide, exam, and PowerPoint presentation to complete a program that technical schools could use.

The following year, Steve spoke at our HVAC Excellence conference and also introduced his soft skills program to our attendees, who were predominantly technical college educators. Our conference committee was extremely impressed with Steve and invited him back as a keynote speaker for our next conference. Steve has spoken at our conferences ever since.

You will find, as you read this book, that Steve's business insight and knowledge of customer behavior transcends what is usual. I am confident that you will reflect hard and long when you read the chapters on empathy, hygiene, editorializing, and listening. This book is enjoyable to read as it delves into the behavioral, psychological, and physiological manifestations of customer interactions.

Steve makes learning fun. He tells stories and shares lively illustrations that readers will relate to and learn from. I can promise you that when you finish this book, you will be a far different service professional than you were at the start.

Jerry Weiss
Executive Director
HVAC Excellence

Preface

My customer service career began in October 1970, when I landed a part-time job delivering groceries for a local market. Entering the homes of our customers was neither difficult nor uncomfortable—all I had to do was be myself. My parents had taught me about respect, serving others, and honesty, and the lessons stuck.

Serving customers during those early years was an education in itself. I learned that people come in different shapes, sizes, and dispositions. The customers who anticipated my arrival welcomed me into their home and prolonged my delivery with chitchat. Only a few customers thought I was unfit to enter their abode—or, they perceived me as an interruption and stopped me at their front door, grabbed the groceries, and sent me on my way. Either customer disposition suited me just fine because my primary concern was that orders were accurate, delivered on time, and in good condition. The most profound lesson I learned was twofold: always maintain a positive attitude, and establish a stable service infrastructure.

The inception of this book began in southeastern Pennsylvania with a speech I delivered to a local group of service company owners, all of whom employed technicians. I spoke about how to deliver exceptional customer service; and, after investing more than 20 years as a service manager, I was fluent in the topic.

The central theme of my speech was that the cost of training is always less than the cost of ignorance. This is especially true in industries where innovation and change are constant amid evolving industry standards and new government regulations. Well-trained technicians usually get it right the first time, thereby minimizing waste and maximizing customer satisfaction. The icing on the cake is when technicians leverage a positive first impression and learn to convey empathy when serving customers.

After the speech, one of the company owners liked my presentation so much that he invited me to speak at his shop so I could train his employees. I arrived at his shop on a Saturday morning and delivered a half-day training session to my client's satisfaction.

I phoned him after a few weeks to ask about the training results, and my client could barely contain his enthusiasm. "One employee sold an additional $15,000 in preventive maintenance programs since your training. We have never done that before," he exclaimed. His excitement became contagious among other local service company owners. I soon learned firsthand that word of mouth is the strongest, and in some cases the most effective, form of advertising.

During the 1980s, anyone providing one-word career advice probably would have said *"computers,"* reasoning that high-tech's pervasive growth into every facet of life would offer stable career opportunities. Parents steered their children away from blue-collar jobs. Becoming a plumber or refrigeration technician just didn't have the appeal of high tech.

Did computer career opportunities flourish back then? You bet. But like the Trojan horse, high tech brought with it an unseen problem. The Internet enabled both portability and the growth of a remote workforce. Computer workers could be overseas rather than down the hallway, and overseas workers did decent work for less money. As more and more companies outsourced computer-related work to overseas operations, the "computer" career path appeared less and less attractive.

Economic factors have resulted in a shift back to blue-collar jobs. Installing a new hot water heater or new kitchen faucet requires a person to show up at your home or office—this is a task you can't send overseas. A plumber must be on-site and in person. This reality has resulted in a new trend of going back to school to learn electrical, plumbing, heating, cooling and air conditioning, and other trade-related careers. Career stability is the priority, and the trades offer one of the best employment options.

But higher trade school enrollment does not necessarily translate into more employment opportunities. Most service company owners have a difficult time hiring qualified technicians. I learned this firsthand during my extensive seminar travel schedule and by speaking with company owners from coast to coast. Service company owners cite a deficiency of soft skills (attitude, appearance, and communication skills) as one of the top two factors that diminish a person's employability; the other factor is a lack of field experience. Between these two factors, soft skills disproportionately affect the potential employer's first impression of a job candidate. Most service company owners would risk hiring a job candidate who lacks field experience, providing his or her soft skills have created a positive first impression. Chapter 1 of this book explains why first impressions most often result in a final decision.

My own experiences as a service manager for a consumer electronics manufacturing company were similar. I learned that hiring technicians with the most positive attitudes was the safest bet, regardless of field experience. People with positive attitudes learn that each problem they overcome today prepares them for the obstacles lying ahead. The cumulative result of resolving daily problems eventually translates into a lifetime of field experience. Technical skills can be taught, and a person with a positive attitude is more resilient, adaptable, and open to new ideas.

Service companies rely on advertising and word-of-mouth referrals from satisfied customers to grow their business. From a sales and marketing perspective, a technician's positive attitude helps drive these referrals. A positive attitude is free, whereas advertising can be expensive. Adams Hudson, the trade industry's leading marketing expert, says: "Advertising can generate leads and marketing can make you look like heaven's gift to contracting. Yet if your technician's attitude is poor then that is the lasting impression to a customer, effectively erasing your marketing benefit."

A technician's positive attitude has greater alignment to a customer's positive expectation, and this situation renders a customer cooperative and calm. This alignment, along with customer biases, assumptions, and perceptions, is covered in subsequent chapters of this book.

A positive attitude almost always trumps technical ability, especially when you consider tenacity and perseverance as attitudinal factors. I learned first-hand about this reality when I administered a factory recall for a consumer electronics product. The mental strain of being continually blamed for a defective product you personally didn't design or manufacture is exhausting. Some customers become irate, hypercritical, and irrational while others seek assistance from their state attorney general, Better Business Bureau, or other consumer advocacy agencies.

My service technicians had to endure multiple failures and denial from the engineers who conceived the product. As the service manager, I focused on the customers, captured facts and details, and remained constructive in the face of a pervasive quality problem. During the first few months of this product's release, my service technicians were feeling demoralized, and the customers were becoming increasingly impatient as the product's failure rate exceeded 45 percent.

Knowing that what wasn't killing me was making me stronger was not foremost in my mind. My physical and emotional response to the factory recall resulted in a lack of perspective, a decrease in restful sleep, and a constant drive to resolve the problem. Knowing that naming the problem was a prerequisite to resolving it, I settled on a clear, concise, and comprehensive title for the malady: stress.

And thus began a search for every book I could read about stress. The topic of stress overlapped with those of psychology and mental fitness, so this led to even more constructive research. Compiling this newfound knowledge into an in-house training program enabled my service technicians to raise their own stress threshold and remain constructive in the midst of chaos.

Eventually, the engineers implemented a complete and improved redesign of the electronics product, and the problems subsided. In retrospect, my service team was able to endure the hardships because of tenacity and perseverance, not technical ability. Willingness to go the extra mile, maintain a positive attitude, and remain constructive are vital, world-class behaviors.

This book is based largely on my own experiences as well as those of hundreds of clients and seminar participants who have shared their difficulties and triumphs. Many of my own mental habits were developed the old-fashioned way—by making mistakes and figuring out how not to do them again. My goal for this book is to help technicians develop the mental habits that allow them to achieve greater career success.

Steve Coscia

President

Coscia Communications Inc.

www.coscia.com I 610-853-9836

Acknowledgments

In writing the *Trade Technician's Soft Skills Manual*, I prevailed upon many friends and colleagues for inspiration and research and to critique various drafts of the manuscript. Robin Kertis has been editing my *Eastern Pennsylvania Business Journal* columns for many years. I'm grateful I was able to enlist Robin's help in editing the early drafts of this book. Robin's diligence and expertise continue to make me look good in print.

Tom Tracy is the artist who illustrated many of the figures in this book. He has the magical and uncanny ability to draw feelings and facial expressions. Tom has been illustrating my ideas for more than 20 years. I am indebted to Tom's creativity along with his willingness to meet tight deadlines.

Special thanks to my wife, Veronica H. Coscia, and our son, Michael Coscia, who were generous with their time, love, and patience throughout this project.

While there are many friends and mentors who have helped me through the years as a speaker, writer, and consultant, I especially want to thank the following trade professionals who contributed their time, loyalty, and friendship.

Bill Ronayne (www.bvhvac.com) is one of the most bighearted people I know and the president of Brandywine Valley Heating & Air Conditioning Inc. He gave me full access to his company and his technicians so I could learn and observe world-class behaviors.

Warren and Patricia Lupson (www.acca-ncc.org) were among my first friends in the trades. Their commitment to excellence and benevolent attitude toward others serve as terrific leadership examples.

My colleagues and fellow founding members of the Contractor Consultants of America are encouraging and unfailingly wise. Their collective business expertise is in a class of its own. The Contractor Consultants of America founding members are

Drew Cameron (www.hvacsellutions.com)

Joe Crisara (www.contractorselling.com)

Adams Hudson (www.hudsonink.com)

Brian Kraff (www.markethardware.com)

Tom Peric (www.thegalileo.com)

The generosity and thoughtful open-mindedness of Jerry Weiss, Howard Weiss, and Tom Tebbe from HVAC Excellence (www.hvacexcellence.org) led to my relationship with Jim DeVoe at Cengage and the publication of this book. Their collective vision and service to the trades is inspirational.

Art Penchansky (www.hvacinsider.com) has always been gracious in making time to review my company's news releases and articles for *HVAC Insider* Keystone/Garden and Mid-Atlantic editions. His due diligence and follow-up demonstrate his pursuit of the mutually beneficial relationship.

Bob Frank and Gary Bower at American Air Distributing, Inc. (www.americanairdist.com) are an endless source of knowledge, inspiration, and kindness. Their intense professionalism and service to the trades is surpassed only by their audacity to embrace change and win.

Mike Weil and Terry McIver (www.contractingbusiness.com) were two of the first trade industry editors to show an interest in my articles. Their friendship and extraordinary kindness have helped me and many others in the trades.

The Air Conditioning Contractors of America, Delaware Valley Chapter (www.dvcacca.com), hosted my first customer service seminar to the trades. Special thanks to Susan Morris and Steve Dragon for their forethought.

Mike Murphy, editor in chief at *Air Conditioning Heating Refrigeration News* (www.achrnews.com), is a helpful and considerate professional. I am thankful to be counted as one who has benefited from his thoughtfulness over the years.

Special thanks to Tom Jaenicke from Atomik Creative Solutions for introducing me to Jennifer Tomb at the National Propane Gas Association (www.npga.org). The National Propane Gas Association is a world-class organization comprising members who are relentless in their pursuit for quality. The association continually challenges me to reach a higher bar, for which I am most grateful.

I am thankful to the Air Conditioning Contractors of America, Ohio Chapter (www.accohio.org). Its members were among the first to invite me as a speaker at their meetings. Special thanks to Sandy Pogan, Tim Volpone, Scott Robinson, and Gary Jackson.

Halos and Hygiene

The Halo Effect

Andrew has been out in the field on residential repair calls since 7:00 a.m. Just before lunchtime, he receives an emergency dispatch from the shop regarding a broken water heater. He doesn't want to take this call, but the dispatcher says that he is the closest, and the customer is frantic. While heading toward the emergency, Andrew lights up another cigarette. He parks his truck in front of the customer's home, grabs his tool box, and walks toward the sidewalk—where he stops, drops his cigarette, stomps it with his foot, and then approaches the front door. Andrew is hungry and tired; he wants to get this job done quickly.

Mrs. Walker is the customer who phoned in the emergency repair for her water heater. She has been waiting anxiously by the window for the plumber's truck to arrive. She sighs with relief when she sees Andrew's truck. However, her relief quickly turns to disappointment when she sees the repairman leave a cigarette butt on her sidewalk. "How rude!" she thinks to herself.

When she answers the front door, Andrew's rushed manner further heightens her disappointment. Mrs. Walker thinks to herself,

"*If this weren't an emergency, I'd ask the shop to send another repairman.*" *Mrs. Walker is certain of one thing: She will never call Andrew's company again.*

In this instance, Mrs. Walker was disturbed by behavior that had nothing to do with Andrew's technical ability. In fact, Andrew is an ace technician, but this does not matter to Mrs. Walker because of her initial negative impression. It gets worse. Mrs. Walker will later complain to her neighbors about Andrew, further damaging his company's reputation and diminishing future revenue potential in that neighborhood.

FIGURE 1-1

Mrs. Walker sighs with relief when his truck arrives.

Her relief turns to disappointment when she sees him leave a cigarette butt on her sidewalk.

Cigarette smoking used to be more pervasive and socially acceptable than it is today. Eventual release of scientific data regarding nicotine's addictive nature in addition to the health dangers of second-hand smoke has transformed public opinion against cigarettes. Therefore, a negative first impression associated with cigarettes is almost a foregone conclusion—especially among residential customers. Cigarette smoke odor lingers in clothing and is impossible to hide, and therefore customers will react negatively. A few of the many bad side effects of the nicotine in cigarettes on dental health and hygiene are a visible discoloration of the teeth and bad breath.

The first impression is a culmination of the visual and verbal behaviors a service professional transmits. In the opening story, Mrs. Walker behaved the way most people do—she made a quick judgment about Andrew's character based on how he disposed of his cigarette.

Customers may jump to a false conclusion or stereotype a service professional and then seek evidence to justify their decision. People search for justification to verify their biases, which are not always based on fact.

Mrs. Walker didn't like the cigarette, so therefore she didn't like Andrew either. Seeking to justify her bias, Mrs. Walker negatively perceives Andrew's rushed manner as an attitude problem. Andrew has two strikes against him before he can demonstrate his technical proficiency—and

"A halo effect occurs when one or more positive characteristics about a person dominate the way that person is viewed by others."

Mrs. Walker, seeking further bias justification, will likely be ultracritical and high maintenance. Andrew will then perceive Mrs. Walker's negative attitude, and the service call is doomed before it ever started—all because of perception, not reality.

Other behaviors that may result in a negative first impression include a customer seeing debris falling out of a truck (empty coffee cups, food wrappers, water bottles, etc.). It won't matter if a service professional picks up the debris and throws it back inside the truck; a customer's first impression will be, "What a slob!"

The first impression takes only a second or two to form, and service professionals rarely get a second chance to alter the customer's frame of mind. For example, a disinterested or perfunctory technician can transmit a message of, "I really don't care whether you ever call us again." To take it a step further; what impression does a twentysomething technician's nose, mouth, and tongue piercings have on the customer who just happens to be a baby boomer? Technicians may transmit messages they do not intend. But once this happens, it's too late.

A customer's perception of a technician is a highly subjective encounter, because people see things through a different lens or filter. Quality-focused contractors understand this dynamic and train their employees to maximize a positive first impression.

Psychological evidence indicates that we experience our feelings toward someone a split second before we can intellectualize about them. The positive feeling we experience is called a halo effect.

A halo effect occurs when one or more positive characteristics about a person dominate the way others view that person. Therefore, nonverbal behaviors such as facial expressions, body language, and hand gestures translate much about how a technician will be perceived by our customers.

The opposite of the halo effect is called the horn effect. The horn effect occurs when a technician allows one negative aspect of his behavior to influence everything else. Andrew, in the preceding story, wasn't aware that stomping on his cigarette in view of Mrs. Walker was a manifestation of his horn effect. Mrs. Walker has a license to make incorrect assumptions because she is the customer (the customer is always right). Therefore, technicians must practice a very simple and profound rule of conduct. The rule is, "Customers won't be able to react to the horn effect if they don't see the problem in the first place." Just don't do it.

Both halo and horn effects can be conveyed in rather subtle mannerisms. Nonverbal behavior plays a big part in a technician's performance. Similar to any behavioral transgression, some are sins of commission and others are sins of omission. Andrew's sloppy cigarette butt stomp was clearly a sin of commission. The absence of a smile and a polite greeting are sins of omission. Not wearing shoe covers inside a customer's home is another sin of omission.

When a technician learns to maximize the halo effect to his or her advantage, then business relationships can flourish. A customer's perception about a technician, whether it's wrong or right, is their reality. The halo effect transcends one-on-one encounters like an interaction between a customer and a service professional. A customer's perception about a service professional (one-to-one ratio) results in the customer assuming that all the employees at a company behave similarly (one-to-many ratio). While the customer's assumption is unfair, it is the reality that service companies face. The mathematical implications of customer perceptions can be profound. Customers will tell others about their dissatisfaction with a service company much more readily than they will share a positive experience.

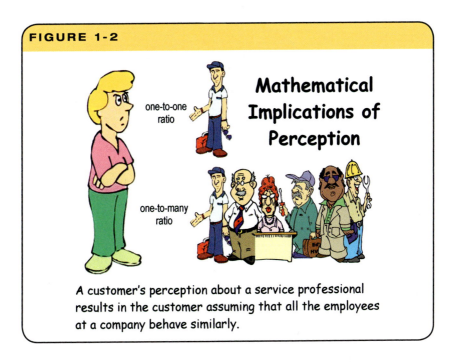

FIGURE 1-2

Mathematical Implications of Perception

one-to-one ratio

one-to-many ratio

A customer's perception about a service professional results in the customer assuming that all the employees at a company behave similarly.

The Internet has accelerated the pace of information sharing with the proliferation of websites devoted to consumer reports and customer satisfaction ratings. Digital information is available at the click of a mouse, and it is shared among thousands of people instantaneously. Averting these events is almost impossible; service professionals must follow up with negative feedback immediately in an attempt to remedy the problem.

The Internet provides cover and anonymity for complaining customers, who will write in a more scathing and extreme manner than they would when speaking face to face. Writing provides a person with a level of separation through which they feel insulated from your presence.

When handling a customer's written complaint, service professionals must practice emotional detachment regarding the customer's written remarks and focus on fixing the problem. The speed with which a service professional responds to a customer's problem has a direct correlation to a customer's behavior. The sooner you respond, the better.

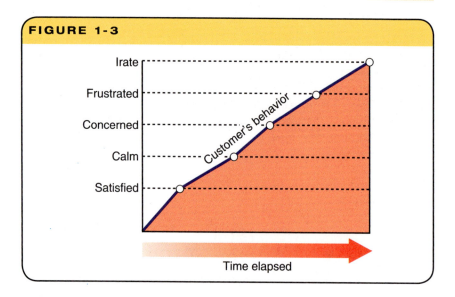

FIGURE 1-3

First and foremost, handle the complaint off-line and out of public view. Resolving a dispute with an unhappy customer for everyone to see is a mistake. A private and neutral environment is required for a rational and factual dialogue. Then do what the customer least expects. Search your records for their phone number and then call the customer with a sincere effort in wanting to make things right. Ask the customer for their side of the story and then listen while the customer vents. You will find the customer's telephone demeanor to be quite positive and in stark contrast to their original written assault. Let the customer finish venting and then ask if there is anything you might need to know. When the customer is finished venting, summarize what you heard to clarify the details and then begin resolving the problem. Being sincere and empathetic works wonders with complaining customers.

When the problem is resolved, ask the customer if he or she would kindly revise or delete their online complaint to reflect a mutually beneficial outcome. Never say or write anything off-line that you would not want made public. Just because you have taken a customer's situation off-line does not mean the customer can't publish something new. Therefore, maintain a high level of integrity and professionalism when serving customers.

Hygiene and Neatness

Mr. Martin answered the door to greet Stanley, the Action Heating and Air-conditioning technician who was on-site to perform preventive maintenance. As he opened the door, Mr. Martin was pleased to see Stanley's smiling face; but when Stanley stepped into the entrance to introduce himself, Mr. Martin had to back away due to Stanley's "coffee breath." The coffee and doughnuts that Stanley had just devoured while sitting in his truck had left a pervasive and pungent odor that Mr. Martin found offensive.

"When a customer is finished venting, summarize what you heard to clarify the details and then begin resolving the problem."

Personal hygiene elicits information by which customers will judge a service company. Stanley's smile and polite greeting helped to reinforce his halo effect until his coffee breath changed everything. The obvious neatness factors, such as wearing clean, well-pressed company shirts, are a given; however, when we venture into the topic of personal hygiene, things become a little more visceral. Perhaps it's because personal hygiene involves our sense of smell, and therein lies the problem. Why? Because our brain handles the sense of smell separately from our other senses.

FIGURE 1-4

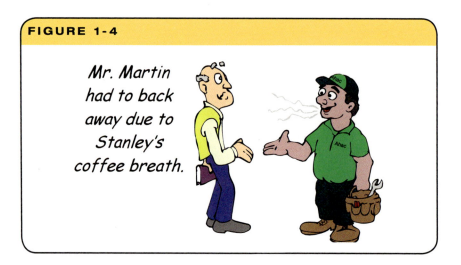

Mr. Martin had to back away due to Stanley's coffee breath.

Our brain's thalamus (Latin for "inner chamber," referring to its position in the brain) manages input from all the senses—except for the sense of smell. What we hear, see, taste, and touch is handled in one place (the thalamus), and the sense of smell is handled in another—the prefrontal cortex. But why is this so important? The prefrontal cortex is also where our brain stores memories, along with our emotional response to what we sniff. There is a link between smells and our memories and our emotions—a rather explosive combination of entities.

Perhaps that is why odor evokes such an emotional response from people. And it's also why events that involve pungent odors are hard to forget— our memory associates odors with events and people. Residential service professionals have probably entered homes that possessed a foul odor. Will they ever forget the location of those houses? Probably not—because smell is linked to their memory. Similarly, an adult who opens a can of Play-Doh and takes a sniff is almost immediately filled with a rush of childhood memories and emotions. The triple threat of smell, emotion, and memory can elicit an extreme response. Needless to say, customers will never forget the technician with coffee breath or body odor.

The halo effect and horn effect rely so heavily upon subtle factors of customer perception. The best way to avert conveying a negative impression is to not get there in the first place. Always keep a fresh shirt in the truck, along with breath mints, mouthwash, and premoistened wipes.

INSIGHT

Examples of the halo effect and its impact on market dominance, sales momentum, and customer acceptance make robust business case studies. The Apple iPod MP3 player and its meteoric rise and acceptance among consumers was due to the halo effect. Customers perceived the Apple iPod to be new and innovative, and these perceptions, like most biases, were not based on evidence. Yet most mainstream customers assume that Apple invented the MP3 player—this is not true. The first Apple iPod appeared in retail stores on November 10, 2001, more than a year after the first mass storage MP3 player hit the market.

Historically, mass storage MP3 players were the inception of a Singapore company called Creative Technologies Limited, a leader in the digital audio field. In April 1999, Creative Technologies launched the NOMAD line of digital audio players, which included an MP3 player with a 6 MB hard drive. It was called the Jukebox. By the late 1990s the miniaturization of nonproprietary hard drives, flat-panel displays, and integrated circuits paved the way for the MP3 player. Using nonproprietary off-the-shelf parts, Creative Technologies introduced the first mass storage MP3 player to the consumer electronics market. The idea of having your entire CD collection loaded onto one device was radical—a quantum leap.

Early innovators of radical ideas suffer from the daunting task of breaking through the old paradigm, creating awareness, and overcoming customer confusion and skepticism about a quantum leap in technology. Creative Technologies' attempt to break new ground and overcome confusion was accomplished by designing their NOMAD Jukebox to resemble its forerunner, a portable CD player. The brushed metal; mostly round case, just slightly larger than a CD disc; and black earphone cable would hopefully give consumer electronics customers an easier time relating to this new innovation. Among audio enthusiast geeks, the NOMAD Jukebox was a breakthrough product.

Apple, looking to break into the consumer electronics audio market, used the same nonproprietary off-the-shelf components and introduced its iPod. The iPod's milky white design was rectangular in shape, about the size of a deck of cards, and had a matching white audio cable. Apple's simple yet elegant design had no resemblance or relevance to a portable CD player, which resulted in a halo effect among mainstream customers. The positive vibe was based on feeling, even though the iPod was made from similar components as its predecessors. Customers assumed the iPod was new technology based on feelings, not facts. Within three years, the iPod's popularity and exponential sales resulted in a 90 percent market share of the hard-drive-based MP3 players.

The iPod assumption among mainstream customers was based on a mental shortcut similar to most biases. It was the halo effect that catapulted the iPod to mass acceptance.

A little precaution can go a long way. To venture even deeper into the topic of personal hygiene, I would be remiss if I did not mention other offensive odors such as perspiration, foot odor, flatulence, and even dirty clothing. These socially unacceptable odors may seem extreme, but they do occur.

Service managers are urged to be discreet when addressing this issue with employees. Being discreet means meeting privately and in an inconspicuous manner to draw less attention to the meeting. My suggestion to have a discreet meeting creates a mutual awareness and the eventual resolution to this problem. In all problematic situations, I advise professionals to practice containment—this means not making things worse, and this chapter will suffice as a reminder to technicians that their customer's sense of smell will create a memorable perception about your company's service.

First Impressions

During a training session with a residential service company, I spoke about the importance of first impressions. The employees and I discussed numerous nonverbal scenarios when a customer service professional conveys a message. The usual body language and facial expression situations arose, along with lots of laughter. Then things became a little more personal. One of the attendees, a technician named Jim, rolled up his sleeves to reveal several tattoos on both of his arms. I invited Jim to the front of the class, and he eagerly opened the top few buttons on his shirt to give the class a partial view of more tattoos on his chest.

Needless to say, Jim was heavily tattooed and very comfortable with his appearance. In addition, he was tall, smart, and articulate. He had a great smile, a positive attitude, and a muscular build. Jim's verbal presentation was in stark contrast to his physical appearance. This is because a big guy with tattoos conveys a stereotypically negative image of a troublemaker.

Jim admitted that this had been a problem for him in the past. "Customers were afraid of me when I wore sleeveless shirts," he said. "When I asked a customer a question, it was hard for me to break through their first impression. Their visual image of my tattoos had already created a bias, so that even though I spoke in a polite and professional manner, it didn't matter. The customer's mind was already made up."

Jim knew his employer trusted him to behave responsibly; so without being asked, Jim used common sense and averted future, negative first impressions by wearing a long-sleeved shirt.

In Jim's case, a manager did not have to intervene. However, what if Jim was reluctant to wear a long-sleeved shirt? Would a manager have the right to insist upon a dress code in the best interest of ensuring customer satisfaction? The answer is yes, providing the company has set expectations and a published dress code.

Company owners must enforce a dress code and a positive appearance that is congruent with a company's image. Your customer's bias, right or wrong, must be measured and used to influence a company's customer service demeanor.

"Always keep a fresh shirt in the truck, along with breath mints, mouthwash, and premoistened wipes. A little precaution can go a long way."

The best method for beginning an objective dialogue is by asking a simple question, "What is best for the customer?" The answer to this question must drive a company's effort toward conveying the best first impression.

Summary

Here are a few principles that service professionals can employ to help convey a more positive first impression:

- Be mindful of the first impression and the halo effect.
- Keep your truck and workspace neat and orderly.
- Don't leave debris on a customer's property.
- Practice good personal hygiene before greeting a customer to cover foul odors and bad breath.
- Wear clean clothes or company-issued clothing.
- Wear shoe covers when entering a customer's home—use drop cloths when carrying equipment in and out of the customer's home.
- Clean up after yourself when the work is done. Respect a customer's home.
- After your work is done, show the customer what you did in very simple terms. Let the customer see the evidence of your valuable service.
- Answer any questions posed by the customer.
- Conclude each service event by saying something like, "Thank you for doing business with us."

As you can see, the behaviors just described have a positive impact on the first impression. Business owners have an overwhelming responsibility of navigating their organizations through a challenging and sometimes uncertain environment of managing employees.

Company owners and managers can enforce policies for dress code, appearances, and neatness. These policies should be clearly explained both verbally and in print. Employees, and their managers, should sign this policy agreement for future reference in the event violations occur. Policy agreement should be kept in the personnel file for each employee. In addition, verbal and nonverbal communication procedures must be established.

If left unchecked, the "people problems" in organizations will cause you to squander your time and resources. This condition causes some managers to retreat rather than lead. Don't let this happen to you. Your employees look to you for leadership. If it is not forthcoming, they will behave in a manner that is most comfortable to them. Unfortunately, this behavior might not be in the best interest of the company.

The Customer Is Always Right

2

Different Greeting Styles

For the last 30 years, Stan has been repairing customers' kitchen appliances. He enjoys his job, and he especially likes people. Stan has learned that residential customers come in many different behavioral styles. Most customers are friendly, and a few are not. Some customers talk too much, while others barely acknowledge his presence. Stan is prepared for anything.

He was reminded about different customer styles when he rang Mrs. Smith's doorbell and found her totally preoccupied with a phone call and holding her laptop computer. "I am here to repair the refrigerator." Stan conveyed in his cheerful tone of voice. Mrs. Smith ignored Stan. "The kitchen is in the back." she said, as she waved toward the back of the house and then walked away. Stan just smiled to himself as he walked to the kitchen.

Service professionals who perform residential service must be prepared for every type of customer greeting. Some customers will greet a service professional in a terse manner, and others will be pleasant. Stan knows that when it comes to greetings, the customer is always right—even when a customer misbehaves.

FIGURE 2-1

A customer's greeting style may range from terse to ebullient. Either way, the customer is always right.

The kitchen is in the back.

Over the span of 30 years, Stan has learned to convey his best behavior even when customers do not. Pointing out why a customer's behavior is wrong can exacerbate a situation and result in lost business.

In a residential service transaction, both the customer and the service company agree to specific terms and conditions. The service company may agree to be on time, to be courteous, to be professional, to dress neatly, to resolve problems, and so forth. The customer's only obligation is to pay the service bill. The customer's behavior is not a factor listed among the business terms and conditions. Therefore, service professionals who seek payment will overlook customer misbehavior and treat the customers as though they are right.

Customers' homes are their castles, and their behavior may reflect this belief. Overly protective customers may ask to see a service professional's identification before allowing entry. On the contrary, other customers, like Mrs. Smith in the opening story, care little about who arrives to repair their kitchen appliances. In Mrs. Smith's mind, a technician is nothing more than an interruption for her to endure. Customers come in many differ-ent styles, and field service professionals must be prepared with a positive greeting—regardless of the customer's demeanor.

Whether a residential customer greets a service professional with a warm smile or a cold frown, the service professional must always maintain a posi-tive attitude. A service professional's attitude will affect his or her aptitude or ability.

Body language such as facial expressions and posture play a significant role in exhibiting attitude. The demands of serving residential customers require a strong balance between attitude and aptitude, especially dur-ing busy seasons. Service professionals who maintain a positive attitude enable themselves to perform at peak aptitude. When a service professional becomes unnerved by a difficult or terse customer, attitude is apt to suffer. When the attitude goes south, the aptitude is usually not far behind.

FIGURE 2-2

Start Your Day on a Positive Note

Choosing to start the day with a positive attitude is important. This is because what happens in the morning can set the stage for the remainder of the day. Practicing positive expectancy is another vital morning behavior. We create our own self-fulfilling prophecy when we expect good things to happen.

Future success depends on daily, consistent, and positive action. In time, optimistic feelings and positive expectancy will produce outstanding results. Attitude is a choice.

Service professionals who begin their day in an organized and orderly manner are more apt to remain positive throughout their day. It's important to meet with a dispatcher early to clarify the day's schedule and discuss potential travel obstacles. Ensuring that work orders include all the necessary information and that tools and required materials are loaded onto the truck is a good start—investing a few minutes preparing for a busy day results in greater self-confidence. This confidence is manifested in a more positive attitude and in a professional communication style. People speak with greater certainty and authority when they are confident in their details. On the contrary, people who are ill prepared will reveal their lack of preparation and doubt in their body language and tone of voice.

If problems arise, it is vital for a service professional to remain calm and not choose anger or frustration. Not only is becoming angry perceived as poor customer service, it is also self-defeating. When a service professional chooses to become angry, this leads to a downward spiral of diminishing returns.

Another downside of choosing anger involves punishment. Not the punishment you might expect from your boss, but rather the self-inflicted kind. People are not punished for their anger; they are punished by their anger. The downside of anger is a lack of mental clarity, decreased problem-solving skills, and a propensity to become self-absorbed rather than focusing on the customer.

Other people cannot frustrate or annoy us if we choose otherwise. Therefore, the best way to avert anger is to avoid getting there in the first place. Just say no to anger, because anger hinders a service professional's ability to discern and diagnose complex problems.

Smiling and Eye Contact

A consistent and positive greeting should be a standard operating procedure for all residential service visits. When a customer sees a smiling face, the service professional is off to a good start regardless of the customer's frame of mind. Some customers may appear terse or seem disinterested only because they are preoccupied with something else at the moment.

The customer's behavior is outside of a service professional's sphere of control. Therefore, it's best to not get upset about what is uncontrollable. Instead, service professionals should focus on what is within their sphere of control—their own attitude and choices.

An apathetic or disinterested customer should not diminish a service professional's positive attitude and warm greeting, because these culminate in a first impression by which the entire service company is judged. A service professional may not get a second chance to make a positive first impression, so it's best to prepare in advance and begin with a smile.

INSIGHT

A psychology experiment involving facial expressions produced a significant result.

Adult subjects repeated vowel sounds that forced their faces into various expressions. To mimic some of the characteristics of a smile, they made the long "e" sound, which stretches the corners of the mouth outward. Other vowel sounds were also tested, including the long "u," which forces the mouth into a frown.

Subjects reported feeling good after making the long "e" sound and feeling bad after the long "u."

This simple, yet profound experiment reinforces the strong relationship between facial expressions and emotions. The smile makes a person feel good.

A smile is much more than a reflection of our feelings, because a smile also changes the way we feel. Go ahead and smile. How do you feel when you smile? You feel good!

Here are five reasons for why smiling is a good idea:

- Smiling is contagious.
- Smiling makes you attractive.
- Smiling makes you seem successful.
- Smiling helps you stay positive.
- Smiling relieves stress.

When greeting residential customers, making good eye contact contributes to a positive first impression. The eyes are the window to a person's soul. Nonverbal behaviors such as eye contact play a large role in establishing good rapport. Eye contact and facial expressions reveal volumes about a person.

How would you feel if while paying a counter clerk for a few hundred dollars in supplies, he failed to make eye contact? You would most likely feel slighted or poorly served.

The indifference and apathy that we sense from an impersonal counter clerk does little to enhance the value of their employer's business. The lack of eye contact is a subtle way of the counter clerk saying, "We don't care whether you shop here again."

Eyes reveal a person's feelings and customers feel uneasy if they cannot see your eyes. A service professional who avoids eye contact when serving a customer will be viewed negatively. Avoiding eye contact can be considered a sign of dishonesty, lack of self-confidence, and apathy.

The person who maintains proper eye contact is considered warm and friendly, while the person who stares is considered as cold and intruding. Maintaining eye contact does not mean staring at a customer. Staring makes customers uncomfortable. So what's the difference between meaningful eye contact and staring? Socially acceptable eye contact lasts about 3 seconds or so; staring is continued beyond that. Effective nonverbal behaviors convey a willingness to cooperate before any words are spoken. A service professional shares a plethora of information with customers based on his nonverbal demeanor.

For service professionals, eye contact in addition to a smile ought to be a standard operating procedure.

I learned an important lesson about eye contact when I was young. I hitchhiked from Long Island, New York, to Montreal, then to Quebec City and on through New Brunswick to Nova Scotia and then back to New York through New England. It was the summer of 1975, and hitchhiking was vogue among college students who were interested in an economical way to travel and meet students from other countries.

"We invite the type of behavior that we convey toward others."

FIGURE 2-3

The First Impression

When greeting customers, a smile and eye contact contributes to a positive first impression.

One of the first important lessons I learned about successful hitchhiking involved building a relationship with a prospective ride. When I began my hitchhiking trip, I was standing alongside the New York State Thruway. Holding a sign reading "Montreal," I watched as the cars and trucks whisked by me at 60 mph. Upon being stuck in the same spot for more than an hour, with no success and little hope for improvement, I began to assess the situation. After a few minutes of logical thought, I decided to walk to the next entrance ramp and hitchhike from there. My logic was that the vehicles would be traveling at a much slower speed while on the ramp, and therefore stopping would be more convenient. I arrived at the next thruway entrance ramp, and within 5 minutes a driver stopped for me. This driver stopped not only because he was driving at a slower speed, but because he and I had made eye contact that established 2 seconds of rapport.

The effective combination of positive attitude, aptitude, preparation, and nonverbal expressions culminates in powerful customer service behaviors. The oldest and most profound customer service idea is the Golden Rule: treat others the way we wish to be treated. The Golden Rule has its origins in philosophical writings dating from antiquity. I have taken the liberty of paraphrasing the Golden Rule: "We invite the type of behavior that we convey toward others." This interpretation of the Golden Rule emphasizes the cause-and-effect consequences of communication.

FIGURE 2-4

THE CONSTRUCTIVE DIVERSION

Hi! I'm John Landis, the owner of Landis Plumbing and Heating.

Landis Plumbing and Heating

One of my residential customers, Mrs. Edmonds, lives alone in a big house. She has refused my invitation to invest in a service maintenance plan. Instead, Mrs. Edmonds calls only when a repair is necessary.

Mrs. Edmonds' upstairs toilet tank had been filling intermittently. Rather than call Landis for service, she tolerated the intermittency for as long as she could. When the tank basin stopped filling altogether, Mrs. Edmonds phoned Landis for service with the expectation of a quick and inexpensive repair.

Why can't you come right now?

John took a slow and silent, deep breath before responding...

But Mrs. Edmonds, my daily schedule is already filled.

I need you today, not tomorrow! If I needed you tomorrow, then I'd call you tomorrow!

If only Mrs. Edmonds hadn't waited until now, this phone call could have been averted.

John felt his emotions swell, and he almost lost his temper. Then he remembered that **the customer is always right**.

This simple phrase resulted in a more rational approach to how he listened while Mrs. Edmonds complained, and it gave John the inner calm to use a constructive diversion...

Want to save 10% on this toilet repair?

The two seconds of stunned silence on the other end of the phone told John that his constructive diversion had worked.

Mrs. Edmonds' affirmative response indicated her willingness to hear more. This gave John the opportunity to convey the benefits of preventive maintenance. Suddenly, using the downstairs toilet until tomorrow wasn't such a big inconvenience to Mrs. Edmonds.

The Constructive Diversion

Some customers have unreasonable expectations, and their motive is usually not malice or disdain. Instead, unreasonable expectations result from being uninformed. Therefore, the service professional must remain calm, resourceful, and—most important—constructive. The following illustrated story depicts how a constructive diversion can change a customer's attitude with a calm and resourceful reply.

In this story, John used a containment problem-solving method during his phone call with Mrs. Edmonds. Realizing things were not under control, John knew the situation would need to be contained before he could qualify the real issue and then correct the problem. Holding his tongue and remaining calm while Mrs. Edmonds vented her frustration enabled John to conceive a constructive diversion that shifted control back to John.

Mrs. Edmonds was wrong to behave the way she did, and yet John maintained his belief that the customer is always right. How can this contradiction make sense? The answer lies in John's attitude. He has learned that staying focused on the customer's wrong behavior leads to a downward spiral of mutual distrust, emotional outbursts, and an outcome in which no one wins.

Customers are wrong sometimes. They make incorrect assumptions, exaggerate the facts, or become inflexible in their point of view. Regardless of what a customer does wrong, a service professional must maintain a positive attitude.

A moment of truth occurs when preparation and training are put to the test. It is a situation when rational thinking, a calm demeanor, and a deliberate response must prevail. Most important, you have only one chance to get it right. If John had given even a hint of snippy retaliation, or apathy, then his confrontation with Mrs. Edmonds could have escalated. Remaining calm is not easy, but it is achievable with practice and application even in the midst of a personal attack.

What causes some customers to behave so badly? It may be that they have been rewarded in the past for their bad behavior with special treatment or immediate service. These customers have learned the cause-and-effect relationship between their exaggerated behavior and exclusive accommodation. Some believe that being a customer entitles them to be demanding and condescending. And still others either don't think before they act or are just plain ignorant.

The challenge for service professionals is to remain calm, think rationally, and not take things personally when customers imply blame or make their criticism personal. The ability to detach oneself from a situation while preserving the customer's self-esteem is a key part of rational thinking—no matter what. The likelihood of a mutually satisfying resolution increases if a service professional can maintain good composure, respond appropriately, and show genuine empathy for the customer. However, once the

"The likelihood of a mutually satisfying resolution increases if a service professional can maintain good composure, respond appropriately, and show genuine empathy for the customer."

customer's self-esteem is damaged by an inappropriate response, the offended party might become defensive or, worse yet, play the victim with even more reason to carry on.

Emotions also affect service excellence. An emotional interaction will impact customer retention—but such interactions can be avoided if the service person uses tact, empathy, and professionalism. One of the earliest customer satisfaction studies revealed interesting metrics about emotions and customer retention. The study was conducted by the Technical Assistance Research Programs Institute in the early 1980s among customers with a problem involving a product or service worth more than $100. Customers who had the chance to complain about their problem were twice as likely to become repeat customers as were those who didn't complain. The emotional act of venting is therapeutic, and it enhances rapport. When customers sense that their complaint is understood by a service professional who conveys a little empathy, then mutual trust ensues. People need to be understood to feel valued and appreciated. Therefore, allowing customers to vent their frustrations can be a constructive interaction.

Keeping the Door Open—When the Customer Is Wrong

Customers have a license to misbehave. A service professional who understands that the customer is always right knows that the behavior is not the real problem. A customer's bad attitude and subsequent bad behavior shouts the assertion, "It's all about me!"

As one-sided or unfair as the "me" stance may seem, it is a reality that service professionals must face. Therefore, a company's service culture should reinforce service skills to ensure that customers feel like it's all about them. If a customer feels good about your service delivery, they're more likely to buy from you again in the future.

Let's examine a specific customer behavior that exhibits the "It's all about me" attitude—in this case, the routine is price shopping. A customer's practice of shopping for the best price is common in home service industries such as HVAC and plumbing.

What usually happens is that a customer needs a plumber to fix a broken faucet and, in seeking the lowest quote, calls many plumbers and heating companies. The customer's "It's all about me" attitude drives this behavior. It's all about getting the lowest price. The customer is not interested in a business relationship; instead, he is seeking a commoditized service, performed at the lowest cost because it is assumed all heating and plumbing companies are the same.

During the customer's quest for a cheap price, the feelings of the responding phone reps at local plumbing and heating companies are of no consequence. Customers are focusing on price.

"How much do you charge to fix a broken faucet?" asks the customer in a terse and perfunctory manner. If the phone rep begins to check for symptomatic details or asks when the faucet started leaking, the customer may curtly interrupt the phone rep by saying, "Just tell me how much you charge."

Anyone in the plumbing and heating business understands that quoting prices over the phone is a precarious practice. Without vital details and visual confirmation, there's a good chance a price quote will be incorrect. But customers still demand a price as a result of their "It's all about me" attitude. So, if the customer doesn't receive a quote, he feels justified in calling another plumber until he gets what he's looking for. The customer ends any hope of a business relationship and slams the proverbial door shut. Any hope of reopening the door is remote due to the customer's focus on price.

FIGURE 2-5

Keep The Door Open

The icing on the cake occurs when a skilled service professional invites a customer to call back after searching for the cheapest guy in town. This invitation is known as keeping the door open, and it is very effective. Success with the strategy of keeping the door open lies in the reality that a customer's search for the cheapest guy in town often fails. This predicament puts the customer in a difficult and sometimes embarrassing situation, because the leaky faucet still needs to be fixed. So who does the customer call now? The customer calls the home service company that extended the nicest invitation along with a persuasive, value-based explanation.

Furthermore, if he hires the cheapest guy in town, then the customer gets what he paid for. Often, the cheapest guy in town may not be well trained, experienced, or certified; and this can result in a botched job. Again, the customer still needs help—and who will he call? Most likely, he will contact a service company that conveyed the most professional demeanor.

So when a customer asks, "How much do you charge to fix a broken faucet?" the service professional's response should be, "That's a great question, and we can give you a quote after we get visual confirmation about what is wrong. Shall I schedule an appointment for you today? Our trucks are fully stocked, our technicians are all certified, and our work is guaranteed. May I please have your name and street address?"

When a service professional focuses on value rather than price in a well-paced, articulate, and friendly manner, a new dialogue begins. This new dialogue challenges a customer's mind-set about whether a customer should entrust their home and their family's safekeeping to the cheapest guy in town. It's subtle and effective, and in many cases customers will pause and consider a value-based proposition.

In all customer relationships, there exists a door through which a business relationship can emerge; and this idea must become part of a company's service culture. World-class service companies view all customer inquiries as opportunities that require a courteous and constructive response. It is the role of managers and company owners to ensure that their staff is well-trained to deliver a value-based service message that invites customers to call back.

Summary

Here are a few reminders about how to shift and reframe your understanding of the idea that "The customer is always right":

- The first impression begins with a professional greeting, regardless of the customer's behavior.
- Disagreeing with a customer about why their behavior is wrong is a losing game.
- The smile and good eye contact communicate a positive message.
- Start the day with a disciplined routine and diligent preparation.
- When the customer is wrong, maintain a positive attitude.
- Practice restraint and not retaliation when handling difficult customers.

- A well-executed, constructive diversion can shift control back to you.
- Detach your emotions from potentially adverse customer situations.
- Think rationally and stay calm to be more constructive.
- Convey empathy to build more positive rapport with customers.
- Balance expertise with more empathy when explaining technical jargon.
- When customers seek a low price, a service professional should reinforce the benefits and value.

The phrase "The customer is always right" unfortunately has confused some service professionals due to an apparent incongruity between reality and the phrase itself. This confusion might cause some to focus on and attempt to fix a customer's behavior. Service professionals must strive to fix problems, not bad behavior. Getting to the root cause of what is making the customer behave badly will eventually fix the behavior. Therefore, when a customer is wrong, it is the service professional's primary job to contain the situation so it doesn't get worse. The ability to keep a situation contained takes practice, application, resiliency, and a positive attitude, along with a desire to do what is correct.

Customers and Congruency

3

What You See Isn't Always What You Get

When her kitchen light switch emitted a spark, followed by complete darkness in her kitchen, Mary jumped back in a state of alarm. "What now!" she exclaimed. Her 50-year-old home was showing signs of wear, and this latest electrical problem forced Mary into action. During the last two years, she had paid a local handyman to fix and replace electrical outlets and lighting fixtures. He was a nice man with a messy, unmarked truck, and he didn't charge much. And while he fixed what was broken, Mary assumed that he lacked knowledge about how to make recommendations to avert future electrical problems. In her mind, he was nothing more than a fix-it guy.

Mary had the feeling that the electrical system in her house wasn't quite right and that perhaps her situation required a more experienced electrician. Mary searched the Internet for "local electricians" and selected a nearby electrical contracting company named Action Electrical. Her choice was based on the company's website advertisement, which listed many industry certifications and affiliations. The photo in the advertisement featured a new truck with a large company logo painted on the truck's panel. It looked very professional and in great contrast to the fix-it guy's truck. Therefore, Mary assumed, based on her website experience, that a "certified" contractor with clean, well-marked trucks was the "best" contractor.

When Mary called Action Electrical, she was surprised to hear a voice-mail greeting inviting her to leave a message for a return phone call. "Why didn't a certified contracting company have a receptionist?" When she heard the beep, she recorded a message anyway because her electrical needs eclipsed her skepticism.

When the Action Electrical technician phoned Mary to schedule her repair for the following day, she mentioned her old house, the previous electrical problems, and asked him whether he could assess the house's electrical integrity. He assured her that he had the latest diagnostic equipment and industry certifications to provide top-notch service.

Feeling relieved, Mary scheduled the appointment and awaited his arrival on the following morning.

When a truck arrived in front of Mary's home, it looked nothing like the truck on the website advertisement's photo. The truck's panel was blank—just like the fix-it guy's truck. Mary wondered if this was the same contracting company she had researched on the Internet. Looking closer, Mary noticed a mess of papers and food wrappers littering the truck's dashboard.

"What am I getting myself into?" Mary thought to herself as the electrician approached her front door. "Something isn't right here," She thought, and mentally prepared herself to find another contractor. Her skepticism overcame her electrical needs as she prepared to send him away.

FIGURE 3-1

The website featured a large company logo painted on the van's panel, so Mary assumed this was the "best" contractor.

When a van arrived, it didn't look like the van in the website photo. "What am I getting myself into?" she thought.

The problem in this story can be summed up in one word—congruency, or the lack of it. Congruency is a very powerful business force, because when it is absent from service events, the result is usually a confused or skeptical customer. The skeptical customer thinks, "Something isn't right here," and this mind-set makes a customer indifferent and resistant to future cooperation, until hard facts can prove otherwise.

FIGURE 3-2

Definition of *Congruency*

Noun: To be in agreement, harmony, or conformity.

Mary would have been less skeptical if more **congruency** *had existed between the website advertisement and her actual experience.*

Adjective: Corresponding in character or kind.

Customers are more likely to cooperate when they experience **congruent** *service events.*

For most service companies, scheduling a residential appointment, similar to Mary's situation, requires numerous interactions or events. It may begin when the customer searches online, sees a newspaper advertisement, or searches the yellow pages. This primary event begins a sequence of steps during which a customer begins passively comparing the congruency among these various interactions. When a subsequent step meets the customer's expectation, congruency exists; and this renders a customer cooperative and calm.

The customer's passive comparison becomes active only when an incongruent event emerges because of unmet expectations. The incongruity between the expectation and the actual event transforms passive cooperation into active skepticism. Hence Mary's frame of mind as she prepared to send the technician away.

Customer Assumptions

Mary's story demonstrates the level of biases, assumptions, and perceptions that most customers may possess when seeking a service company. Mary's assumptions about the "fix-it" handyman might have been true; but in the absence of objectivity and verifiable proof, it is unknown whether the handyman possessed the knowledge, skill, and ability to provide more value beyond fixing and replacing electrical outlets and fixtures. Mary's limited experience in working with electrical contractors became an incomplete frame of reference due to the subjective filter through which she made judgments.

FIGURE 3-3

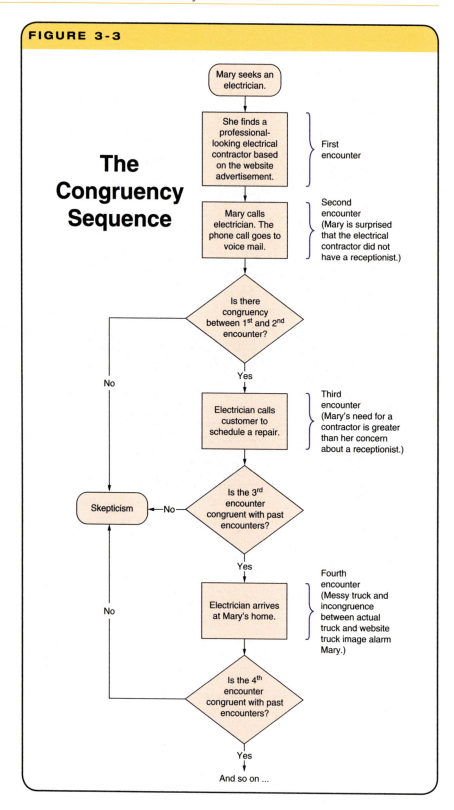

The
Congruency
Sequence

Mary seeks an electrician.

She finds a professional-looking electrical contractor based on the website advertisement.
First encounter

Mary calls electrician. The phone call goes to voice mail.
Second encounter (Mary is surprised that the electrical contractor did not have a receptionist.)

Is there congruency between 1st and 2nd encounter?

No

Yes

Electrician calls customer to schedule a repair.
Third encounter (Mary's need for a contractor is greater than her concern about a receptionist.)

Is the 3rd encounter congruent with past encounters?

Skepticism ← No

Yes

Electrician arrives at Mary's home.
Fourth encounter (Messy truck and incongruence between actual truck and website truck image alarm Mary.)

No

Is the 4th encounter congruent with past encounters?

Yes

And so on ...

"A primary barrier to total objectivity is the 'ego bias.' The ego bias drives people to think, 'Yes, I understand objective versus subjective bias, but this applies to other people, not to me. I'm a very rational and disciplined person with no biases whatsoever.' The ego bias is a form of self-delusion."

Mary is not that different from the rest of us. People often assume traits about others and make decisions based on their own biases rather than on verifiable data. A bias is the tendency people have to judge others or make assumptions based on our own internal filter, which is a culmination of beliefs, experiences, and lessons learned. We all have a different filter, which can result in a subjective outcome. This is why objectivity is such an important factor when working with others.

The difference between objective and subjective thinking is critical. Objectivity means you have no opinion one way or another, and you are not influenced by previous experiences. Your decision is verifiable with facts and proof. On the other hand, subjectivity has a basis in reality, but it is influenced by a person's feelings and experiences, and therefore cannot always be verified with facts and proof.

Most people cannot totally overcome the effect of their biases. This is because our beliefs, experiences, and lessons learned are deeply embedded into our uniqueness and personality. A primary barrier to total objectivity is the "ego bias." The ego bias drives people to think, "Yes, I understand objective versus subjective bias, but this applies to other people, not to me. I'm a very rational and disciplined person with no biases whatsoever." The ego bias is a form of self-delusion.

Mary's bias and key misconception was the assumption that "bigger is better." She assumed that a bigger and more successful-looking contracting company was the better choice. In the absence of verifiable data, Mary's assumption had some merit. In psychological terms, Mary's ability to assume is called judgmental heuristics. This psychological behavior is more common than we might think. People make judgments based on their first impression of someone or something else.

The word *heuristic* means "to discover, to reveal or to invent." Therefore, what a person discovers (based on their bias) during a first encounter with someone else results in their first impression. Mary's quick assumption was based on simplified thinking, which works some of the time; however, it leaves her open to mistakes every now and then.

Judgmental heuristics is a mental shortcut. It's a rule-of-thumb method to shorten decision-making time, thereby allowing a person to function without constantly stopping to think about the next course of action.

In Mary's case, it is easier for her to assume that Action Electrical is the bigger and better company rather than to invest her time in the due diligence of research and fact finding.

Active Skepticism and a Proactive Stance

Consider the electrical technician who is approaching Mary's front door. His positive frame of mind may be focused on the business opportunity of imminent customer satisfaction and of gaining a new customer for future revenue potential. Do you think he is ready for the shock and disbelief of being sent away? The explosive mixture of customer bias, unmet expectations, and incongruence can exacerbate a routine service call. Mary's story continues.

The electrician's smiling face and polite greeting did not sway Mary's skepticism. "I don't want you here," she exclaimed. "Go away." Shocked and confused, the technician just stood there in stunned silence. "Your truck doesn't resemble the truck in your advertisement, and your dashboard is a mess. You don't even have a name badge to identify yourself. Sorry, but I don't want you in my home."

The electrician began to explain that even though his van looks dirty, it's due to be painted the following week. This explanation fell upon deaf ears. It was too late. Mary had already closed her front door.

Confused, humiliated, and frustrated, he considered what went wrong as he walked back to the van. He sat in his van and thought carefully about why Mary closed the door in his face. "Why would she do that?" he thought to himself. "That was just plain wrong." And then he remembered that the customer is always right. "There has to be a logical reason."

As he drove away, he thought about the customer's concerns. Could he have done anything differently?

The van's imminent paint job could have been explained to the customer over the phone, prior to his arrival – that is, if he had made a pre-visit courtesy call. The dashboard would have taken two minutes to clean up. A name tag would cost a few dollars. He calculated how much time, effort, and money it would have taken to avert this mess. He made a list of the changes he would suggest to his boss to ensure that this would never happen again to any of his co-workers.

As the electrician drove back to his shop, two dominant ideas came to mind—personality and process. He realized that the tendency for a customer to make incorrect assumptions is a manifestation of their personality. Likewise, his own assumption about a customer's value perception was a result of his personality (i.e., his belief that customers were concerned only about his technical ability and cared little about whether his truck was clean, painted, and well organized). The subjective nature of personalities can clash. His company can minimize a customer's mistaken assumptions by ensuring congruency between a customer's expectations and the electrician's service delivery. Not doing this would ruin business relationships.

The electrician was convinced that the personality issues could be resolved with more structured and disciplined processes.

FIGURE 3-4

"I don't want you here," she exclaimed. "Go away."

"Why did she do that? That was just plain wrong." Then he remembered that the customer is always right.

Personality and Process

The small investment of time, money, and effort required in being proactive is in stark contrast to the big loss in reputation and revenue. Minimizing customer confusion with a pre-visit call is a vital process for maintaining positive congruency. In the anecdote, a 2-minute courtesy call before arriving could have averted much adversity. As a process, this pre-arrival phone call yields positive and more predicable outcomes. During a pre-arrival courtesy call, the technician can establish an estimated time of arrival and answer any of the customer's questions.

Customers are calm and cooperative when they know what to expect and when to expect it. In addition, a technician can use the pre-arrival call to ask a few simple questions of his own.

These questions might include:

- Where would you prefer I park my van?
- Is the equipment easily accessible?
- Are there any pets in the household?

The answers to these three questions help minimize confusion and surprises. They establish a mutual understanding as well as a more predictable outcome. In most customer interactions, a small investment of time can yield big results.

Customer service events involve a balance between personality and process. Our emotions, responses, beliefs, and perceptions are a result of our personality. Therefore, a customer service outcome, based on personality, will vary depending on numerous factors. These factors can change daily, and they include schedule, physical health, family issues, co-worker cooperation, the weather, and traffic. These factors are outside our sphere of control, so the outcomes will be unpredictable depending on which factors are dominant. Now imagine these factors multiplied by all the people whom a technician might encounter. Each person has their own personality, which is being affected by their schedule, their physical health, family issues, co-worker cooperation, the weather, and traffic. Each personality is different and yields a different outcome, resulting in cumulative and random events.

Simply stated, processes are predictable; personalities are unpredictable. It's easy to see that a personality-based approach to serving customers is destined to fail eventually.

Complacency and Personality

Technicians get into trouble when they become complacent and rationalize their bad behavior based on personality factors. For example, a technician who feels rushed on a busy day decides to save time by not wearing his shoe covers. His shoes look clean, and wearing the covers will not make a difference. He skips wearing the shoe covers during a residential customer repair, with no consequence. "That wasn't so bad," he thinks; and so he skips shoe covers during his next service visit. Throughout his busy day, with the perceived benefit of time saved, he establishes a new habit of not using the shoe covers, even on days when he is not that busy.

Service professionals who become complacent about standard operating procedure eventually get into trouble. Complacency occurs when after doing something well, due to much discipline and preparation, you see how much less you can do the next time and still achieve the same results. Don't get caught in the complacency trap.

This new habit will eventually get the technician into trouble. It's just a matter of time before he messes up a customer's white carpet by not following standard operating procedure (SOP). The customer's complaint phone call to the technician's boss will consume much time and money. The few minutes of time saved by not using shoe covers will pale in comparison to the business expense of handling an unsatisfied customer.

A process implies stable, consistent, sequential, structured, and predictable outcomes. Process steps should be written for future reference. The best service professionals practice a simple rule about processes. Document what you do; then do what you document. Therefore, if the SOP includes wearing shoe covers, then do it—regardless of how busy the day is. Referring to a written process enables a service professional to stay within the SOP parameters. Following the SOP minimizes the personality factors that can derail a service professional's best efforts.

Congruency and the Community

Neighbors share information about service professionals by word of mouth. A positive testimonial from one trusted neighbor to another is a strong marketing tool. Smart service professionals maximize their halo effect in an effort to be likable and get more referrals. Therefore, building a reputation in the community for being courteous, helpful, and efficient is vital.

"Complacency occurs when after doing something well, due to much discipline and preparation, you see how much less you can do the next time and still achieve the same results."

A service professional's community presence goes beyond their ability to diagnose problems and fix equipment. Traveling from one customer to the next in a van that proudly displays their company's name, logo, and contact information can be powerful advertising—not to mention that when they are on-site, service professionals are advertising to the neighbors who will likely need similar services.

When a service professional is driving in a neighborhood and arrives at a four-way stop sign, it's best to be courteous and allow other vehicles to go first—even if the service professional has the right-of-way. Neighbors will associate a courteous driving gesture with the company's name, logo, and contact information. After being extended the courtesy of going first at an intersection, a local neighbor might think to themselves, "That service company is nice." The neighbor then reasons that all employees at a particular service are courteous based on the actions of one employee. This quick assumption is another profound example of a mental shortcut based on very little factual data—more validation for the halo effect and congruency at a very local level.

In communities where neighbors know each other, word of mouth is a powerful force for either good or bad testimonials. Therefore, service professionals must be courteous and careful when driving in communities to avert the possibility of causing negative word of mouth.

Feeling rushed, harried, and stressed while driving from one job site to the next is the most common reason for vehicular mishaps among service professionals. A well-marked van or truck offers no anonymity. It requires very little effort for another driver to call the phone number on the side of a van and complain about a service professional's driving behavior.

The negative impact of hundreds or thousands of passersby who might see a company van or truck causing a traffic accident is damaging to the company brand.

Mike Richards, an electrician for Action Electrical, was late for his next service call and was driving at just above the speed limit. He was in a rush due to numerous delays while wiring his last customer's electrical circuits.

Linda, the company dispatcher, called Mike to say that his next customer was upset. "Their electrical power has been out all day," exclaimed Linda.

"I'm on my way!" Mike yelled. He could feel his stress level escalate due to the mounting pressures of setbacks, traffic delays, and other unforeseen obstacles.

Not knowing when the upcoming traffic light turned yellow, Mike took a risk and accelerated his van in the hope of averting another delay. The light turned red just before Mike entered the intersection. Mike hit the brakes, but it was too late—the van was going too fast. Exceeding the speed limit, the van screeched into the intersection and hit an oncoming car.

While no one was injured, Mike's ego hurt badly because everyone in town knew Action Electrical. Passersby pointed to the company van and made negative remarks.

FIGURE 3-5

Passersby pointed to the company van and made negative remarks.

Mike's irrational state of mind contributed to his traffic accident. Service professionals must be alert to their own mental state and overcome their own ego bias. Mike would have benefited from a 30-second pause during which he might have considered his priorities along with what was within his sphere of control. In a calmer mental state, Mike would have realized that the wiring and traffic delays were outside of his sphere and thereby not something he could control. Continued rational thought would suggest that perhaps Mike's best reaction should have been the conditions that were within his sphere of control—namely, his driving.

A brief pause, even for a few seconds, gives a service professional more perspective and greater clarity. Choices made after a few seconds of rational thought are almost always more constructive than choices made in the heat of the moment without rational thought.

Summary

Here are a few principles that service professionals can employ to help ensure more congruency in working with customers:

- Congruency is a powerful business force. Customers seek congruency in their encounters with service professionals.

- A lack of congruency results in customer skepticism, when a customer thinks, "Something isn't right here."

- People have biases that can result in a wrong assumption. The ego bias is a form of self-delusion.

- People make judgments based on their first impression of someone or something else.

- Passive activity becomes active when an incongruent event emerges, such as an unmet expectation.

- Be proactive and set up your next visit. Make a phone call to establish an estimated time of arrival (ETA), and ask the customer about equipment accessibility, parking preference, and pets.

- Proper balance between personality and process is important.

- Always follow standard operating procedure (SOP).

- Complacency is a bad habit.

- Service vans and trucks promote their brand in a community. Courteous driving maintains positive congruency.

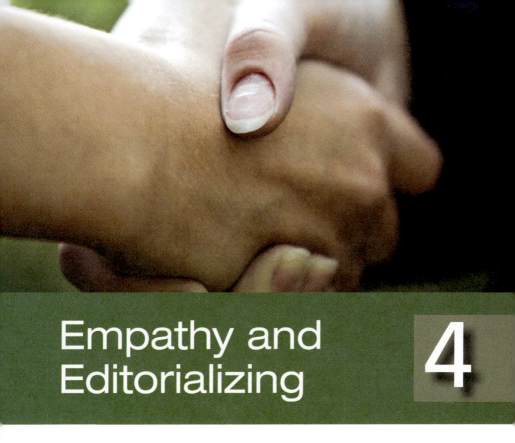

Empathy and Editorializing

4

The Culture Within a Company

We have already learned that customers seek congruency among the sequential service events they encounter. Positive, congruent service events minimize customer concerns and skepticism. The culture within a service company helps achieve consistent and positive congruency among all employees.

Service culture is not established by policies, procedures, and infrastructure alone. It is most often an extension of the company owner's personality.

A company owner's personal qualities—such as persistence, resiliency, patience, and candor—culminate in a unique personality that in turn creates a culture.

Here are a few business examples:

- Thomas J. Watson, the founder of IBM, experienced both successes and failures in his youth. He learned from his failed experiences. These events helped mold him and, by extension, the policies by which he made IBM a global entity.

- Jack Welch's 20 years of employment at GE, prior to becoming GE's CEO in 1981, helped shape his ideas for streamlining the company's bureaucracy. His personal beliefs about a no-nonsense, candid style of management transformed GE into an efficient and vastly profitable market leader.

"Going the extra mile for a customer or coworker is a manifestation of concern and empathy."

- Meg Whitman, eBay's ex-CEO, joined the company when it had only 29 employees. Her professional experiences in working for brick-and-mortar as well as online companies helped prepare her for eBay's exponential growth and success.

A definition of *culture* is "the quality in a person or society that arises from a concern for what is regarded as excellent in arts, letters, manners, scholarly pursuits, etc." I like this definition because it is concise and comprehensive enough to include the key phrase "concern for what is regarded as excellent." If we drill down even further into this phrase, the word *concern* is the clincher. A person's concern for others implies the presence of empathy, and it is empathy that motivates employees to put others' needs ahead of their own. Going the extra mile for a customer or coworker is a manifestation of concern and empathy.

The word *concern* also entails leadership excellence in an emotional and gut-level style of management that employees just can't ignore. The way a leader treats employees, customers, and suppliers becomes a visible example for others to emulate. The "do as I say, not as I do" approach to leadership does not work for the long haul, because it is incongruent to service excellence. Company owners and managers who do not adhere to their internal culture allow others to do likewise because they are leading by example.

The cumulative behavior of all employees (managers, technicians, and office staff) creates a company's culture. Culture involves a shared and learned system of values, beliefs, and attitudes that shape and influence behavior. It is implied that people learn from each other through their shared experiences.

Courageous company leaders keep the performance bar high and have no appetite for mediocrity. The word *leader* by its very nature means that followers exist. Employees working within a culture of excellence lead too, especially in the way they demonstrate an attitude of teamwork and cooperation.

Contractor companies should have one service culture that guides all their internal and external customer interactions. The internal customers are coworkers, including superiors and subordinates. The external customers are those being served and paying for services rendered. Treating internal and external customers with the same courtesy and professionalism is a difficult but worthwhile endeavor.

Maintaining service excellence among internal customers requires strong leadership, especially when unforeseen events arise and burden one department disproportionately. When one department becomes overwhelmed, it is in everyone's best interest to help. Examples of clashes between multiple service cultures include departmental infighting between office staff and field technicians or between finance and sales.

Employees who go the extra mile for their coworkers are leading by example and demonstrating service excellence for other employees to witness.

Culture, Attitude, and an Anniversary

It's Friday, and Tony, a field service technician for a telecommunications company, had planned to end his workday at 3:30 p.m. so that he and his wife could celebrate their 10th wedding anniversary with a special night out. Tony told his boss and coworkers about his anniversary plans, but Mother Nature was not cooperating on that particular day. Strong winds from the night before resulted in major service outages; and by early Friday afternoon, Tony's anniversary plans began to look rather doubtful. Tony was delayed with ongoing emergency service dispatches that kept him working until 7:30 p.m. Finally, while driving home, Tony thought he still might be able to enjoy an evening out with his wife. Then, at 7:40 p.m., a dispatcher alerted him to yet another emergency. Tony immediately responded to the call, finishing up at 9:30 p.m. The request came from an on-call hospital worker who required restoration of their telephone service. Tony understood the vital services provided by hospitals to his area.

When asked about his wedding anniversary plans, Tony's reply was simple. "I have a terrific wife. She understands my job and why I have to work long hours." Tony and his wife celebrated their wedding anniversary on the following evening.

FIGURE 4-1

Tony... sorry about this, but we have an emergency!

Tony's work ethic led him to subordinate his own needs in favor of serving customers first.

Tony's work ethic led him to subordinate his own needs in favor of serving customers first. His flexibility prevented the dilution of critical manpower resources during a peak demand cycle. Someone else might have thought, "Management approved my time off . . . besides, I'm entitled to it . . . let someone else handle the workload." From a company policy perspective, both outcomes are correct. Is one outcome more correct than the other?

Tony's attitude of teamwork and cooperation enable him to lead by example. One employee who sees Tony go the extra mile for customers might be inspired to do likewise. People learn from watching each other. We learn what is correct by observing others and by determining what other people think is correct. There is strength in numbers—we believe a behavior is correct when we see others doing it.

For service professionals, the safety and well-being of customers during peak demand cycles is a reality. A service professional cannot control the weather, but he can maximize what is within his sphere of control when weather extremes arise. Flexibility and patience along with teamwork and cooperation are all within a service professional's sphere of control.

Young or inexperienced service professionals unsure about how to react to peak demand cycles rely on the examples set by more experienced coworkers. This is where a company's culture really matters. New employees are uncertain and more likely to use the example of experienced coworkers to decide how they themselves should perform.

Going the extra mile for customers, while often an individual effort, is more important due to a pervasive attitude within a company—this is the culture. The underlying emotion that drives a culture of teamwork and cooperation, and in this case Tony putting his customer's needs above his own, is empathy.

Why Empathy Is Important

Empathy is the psychological software that allows people to care about those who need help. It is the capacity to feel the emotions of others. The presence of another caring person makes people feel more connected. It is empathy that transcends teamwork and cooperation to a level of service excellence.

Tony, who is the focus of the previous story, has technical expertise in abundance, but these skills alone comprise only half of what is required. The other half is empathy. World-class service professionals must possess the ability to balance their empathy and expertise.

Technical professionals rely on their ability to diagnose and troubleshoot equipment. Given the proper training, schematics, and process-of-elimination methodology, a skilled technician can zero in on a problem's root cause. The self-confidence gained in diagnosing and fixing problems grows with experience; and after many years, this trait affects a technician's communication style. A technical expert speaks with a greater level of certainty and authority because he knows his stuff. This is evident in his tone of voice. Expertise should be tempered with a dose of humility and empathy. Humility enables an expert to believe that listening to the ideas of others is a worthwhile practice.

If expertise and empathy require balance when serving customers, then the same thing applies to self-confidence and ego. *Ego* in this context can be defined as "an inflated feeling of pride and self importance." The most successful service professionals maintain a balance of empathy, expertise, self-confidence, and ego.

"*If expertise and empathy require balance when serving customers, then the same thing applies to self-confidence and ego.*"

Editorializing—Don't Criticize Your Coworkers

Mr. Smith escorted Jim, an HVAC technician, down to the basement and toward the recently installed HVAC unit. A loud whirring noise from the unit had prompted Mr. Smith's request for service. Standing in front of the unit, which was still shiny and new, Jim put down his tool box and proceeded to scan the ductwork. "Whoever installed this ductwork really screwed up," exclaimed Jim. "I never would've installed it like this."

Upon hearing Jim's remark, Mr. Smith was surprised. Why would Jim slam his coworkers on the installation team? Then Mr. Smith assumed that the noise and the shoddy ductwork installation were related, and his thoughts dwelled on the money he had just paid to Jim's company. "How bad is it?" asked Mr. Smith.

"Don't know yet. But it's a good thing I'm here and not the other guys who installed this." said Jim. While Jim's inflated ego was meant to impress Mr. Smith, it had the opposite effect. Mr. Smith excused himself to go upstairs and call Jim's boss to complain about the installation and to demand a refund.

FIGURE 4-2

Jim's expertise and ego eclipsed his empathy—he editorialized. While Jim's assumption about the ductwork might have been correct, he should have been more discreet. Saying less is more.

In this story, Jim committed a quadruple transgression with a toxic mix of mistakes. Jim's expertise and ego eclipsed his empathy for the customer and caused him to editorialize. While Jim's assumption about the duct-work might have been correct, he should have been more discreet. Jim's first impression is prone to be subjective and results in biases due to the absence of factual data. A discrete phone call to his shop to inquire about the ductwork would have added more objectivity to his findings. The old saying about there being two sides to every story applies here. Inviting in-put from an installation technician would have surfaced more details and resulted in a much more constructive explanation to Mr. Smith.

The imbalance between Jim's expertise and empathy, along with the lack of humility, led to his mistake. When a person conveys his expertise in a boastful manner, as Jim did, then the motivation is an attitude of "It's all about me." In customer service encounters, this self-centered approach to sharing information may feel good for the short term, but there is a long-term risk. What feeds our ego may not always be what's best for the customer and our employer. Knowing what *not* to say is just as important as what is said. "Less is more," certainly applies to what a customer hears during customer service situations.

Had Jim balanced his expertise with an equal amount of empathy for the customer, he would have paused, thought rationally about a constructive plan of action, and proceeded in a more discreet manner. Rational thinking

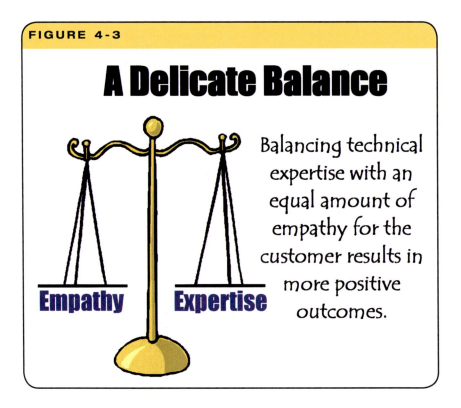

FIGURE 4-3

A Delicate Balance

Empathy   **Expertise**

Balancing technical expertise with an equal amount of empathy for the customer results in more positive outcomes.

would have enabled Jim to be proactive by considering the customer's feelings. Being empathetic implies a genuine concern about the other person. By asking himself, "How might Mr. Smith feel if I mention a possible installation problem?" Jim should find the answer to that question will result in a more empathetic approach to both problem containment and customer satisfaction.

After doing a brief visual assessment of the ductwork and jotting down a few notes, Jim could have gone out to his truck to make a discreet phone call to the shop. In an isolated environment, it's easiest to speak openly, objectively, and candidly. The discreet phone call would enable Jim to emerge from the truck with better information and improved focus on customer satisfaction. The mutual benefits, for both a service company and the customer, increase exponentially when service professionals minimize editorializing.

Conveying empathy is motivated by an attitude of "It's all about the other person." Putting the customer's needs ahead of our own prevents adverse events from becoming worse. Subordinating our own needs puts us in a position to speak less and listen more. When a service professional dominates the listening, then he puts himself in a position to say less and thereby minimize the urge to editorialize.

Editorializing occurs when a service professional shares too much information. This extra information adds no value, and it detracts from customer satisfaction. The worst part about Jim's editorializing is that he tried to make himself look good at the expense of other employees. Among service professionals, this is a major transgression.

In all service, work mistakes will happen. When a service employee finds a mistake, it's best to take responsibility and contain the problem regardless of whose fault it is. In the presence of customers, it is important to be discreet when one service professional finds the mistake of another. The "blame game" just makes matters worse from the customer's viewpoint. Customers want a speedy resolution rather than an excuse. They do not care which employee is at fault. In the customer's mind, the whole company looks bad when employees blame their coworkers.

Editorializing also includes bad-mouthing the competition. Service professionals who criticize the competition in the presence of customers diminish their own image too. It's best to steer clear of this behavior.

A service professional's expertise and ego can cause bad-mouthing, and these editorializing mishaps reveal something to others. Namely, what is revealed is insecurity. Those who chronically criticize others are demonstrating their own constant need for approval. Some technical experts need ongoing approval because they are not satisfied with validating their expertise once—they need to do it again and again. This is an ego issue.

Technical experts who achieve numerous certifications along with the self-confidence and expertise are better off conveying humility. Expertise is best displayed through quiet and mature action that speaks more loudly than words. Let the credentials and experience speak for themselves.

"In the presence of customers, it is important to be discreet when one service professional finds the mistake of another."

When customers ask about "the competition," it's best to reply, "They're a good company. However, may I tell you what makes our company so unique?" Then, with a professional and courteous demeanor, describe your own company's benefits.

Being constructive and averting the tendency to editorialize and bad-mouth others builds greater customer rapport.

The Dr. Jeckle and Mr. Hyde Service Style

Establishing an excellent service culture can be determined by answering a very simple question: What is best for the customer?

The answer is equally simple. Most customers want to be served by professionals who really care. Some service companies do not understand the dynamics and delicate balance between acquiring and retaining their customers. Both customer acquisition and customer retention are the lifeblood of a service company's success. After acquiring a new customer by investing a small fortune in marketing, advertisements, and outreach campaigns, the company's easiest and least expensive practice is to retain its customers. However, some service companies squander money and resources on customer acquisition only to skimp on the resources that will keep those customers coming back. As in *Dr. Jeckle and Mr. Hyde*, the story of a man who transforms himself into a monstrous villain, many service companies maintain dual personalities regarding their treatment of customers.

INSIGHT

How old is your Daughter?

Among the many questions that a young HVAC technician might ask a residential customer first might be, "Where is the thermostat?" or something else related to the equipment. In the customer's mind, there exists a congruity between the question and the purpose of the service call. When the customer's expectation and the service delivery are congruous, the relationship is stable so that rapport can develop.

One question caused grief for Robby, a young technician in one of my recent customer service training sessions. He came to the front of the class and explained that while troubleshooting a customer's thermostat, Robby was distracted by the customer's teenage daughter. He kept his mind on his work. Robby admitted to the class that during this service call, where he felt his

(Contd.)

relationship with the customer was stable, the family's attractive teenage daughter had aroused his curiosity. So he asked, "How old is your daughter?" The customer's subsequent disapproval made it apparent that Robby's relationship with his customer had spiraled downward to instability.

We were all able to laugh about it during the training session; however, Robby made it abundantly clear that there was nothing to laugh about when he returned to the shop. The customer phoned Robby's boss to complain. Customers will make assumptions based on our behavior; and since the customer is always right, we don't stand a chance.

In terms of verbal communication, the errant "daughter" question was an example of editorializing in which too much information made things worse. When service professionals editorialize, they are usually conveying information that contributes nothing to customer satisfaction.

The "less is more" paradigm applies in most business communication, especially when you consider that customers hear what they want to hear. The filters through which customers hear others are based on their biases and assumptions. We all hear things differently, because we all have different filters. And since our biases run deep, it doesn't take much for emotions to escalate when someone crosses the line into socially unacceptable behavior.

Robby was very inexperienced, and this event was a valuable business lesson. The lesson learned was that whenever you're not sure whether you should say the words on the tip of your tongue, then keep quiet. Robby also learned that in business, words can be weapons; and the wrong message can result in a business disaster.

Socially acceptable behavior training, especially in residential service, is a worthwhile investment to avert unintentional blunders. For world class service companies, socially acceptable behavior is a vital customer touch point.

A customer touch point is an event during which a customer comes into contact with your business. Touch points include your website, the logo on the side of your trucks, your front-line phone reps, your promotional materials, and your service technicians in the field, just to name a few. Robby, whether he knew it or not, was a touch point. Congruity across these events is a powerful force in business. Customers will gauge and compare the similarity across each touch point.

Smart service professionals can harness the power of congruity and customer touch points to their advantage as another way to differentiate their service company from the competition.

The Dr. Jeckle and Mr. Hyde service style involves new customers who experience boatloads of gratitude for making such a wise choice. The honeymoon period lasts only as long as it takes for a problem to arise, which results in a call for help. At this juncture, the Mr. Hyde personality of the company may surface—in the form of an apathetic service employee or a mishandled phone call. During these negative encounters, customers can't help thinking about the stark contrast in business behavior between a service company's marketing effort (Dr. Jeckle) and its customer service (Mr. Hyde). This incongruous behavior makes a company's marketing effort appear disingenuous, as though a breach in goodwill has occurred.

Customers disappointed with the Mr. Hyde style of service may defect and seek another service provider. This desertion instigates a customer retention campaign within the service company; and soon the wooing and acquisition, along with the big-budget spending behavior, kicks in to save the day. There seems to be no limit to what a mediocre service company will spend to acquire a customer; but once it has acquired that customer, cost cutting and mediocrity ensue.

The Dr. Jeckle and Mr. Hyde service style is another example of dual and conflicting cultures that waste time, consume resources, and confuse internal and external customers.

Editorializing—Speak Less and Get Busy

Mr. Johnson had scheduled his biannual maintenance for his heating and air equipment. He did this once in the spring and once in the fall to ensure that his equipment would be functioning at optimal level. Historically, the service company sent older, experienced technicians.

Today, however, the service company sent Randy, a young technician who looked like he was still in high school. "Are you old enough to be a technician?" asked Mr. Johnson half jokingly.

"Sure, I'm 20 years old," Randy replied. His smile and positive attitude began to put Mr. Johnson at ease. "Performing equipment maintenance is easy; the older guys taught me how to do it. Plus, I have learned enough in trade school to perform system maintenance. And if I get into trouble, like I did yesterday at a different customer's house, I'll just call the shop and they'll send a more experienced technician to your home to repair the equipment. I still have a few years of trade school before I graduate."

Mr. Johnson began to feel a little less at ease after Randy revealed his educational shortcomings.

As Randy stepped into the foyer, his absent shoe covers were noticed by Mr. Johnson. "The older technicians always wore shoe covers," thought Mr. Johnson. Feeling skeptical about Randy's ability, Mr. Johnson asked "Shouldn't you be wearing shoe covers?"

Randy smiled and answered, "Yeah, I guess I should be wearing them. But my shoes are clean—look for yourself."

Mr. Johnson was sure about two things: Randy talked too much, and Randy lacked an understanding of standard operating procedure.

FIGURE 4-4

Mr. Johnson was sure about two things: Randy talked too much and Randy didn't follow standard operating procedure.

In this story, Randy was off to a good start. He handled Mr. Johnson's first question, about whether he was old enough to be a technician, with a concise and upbeat answer. Randy should have stopped right there. Instead, he chose to editorialize; and this diminished Mr. Johnson's perception of Randy's ability.

When Randy editorialized about the previous day's failed service visit, he was sharing information that Mr. Johnson didn't need to know. In addition, Randy's failure to wear shoe covers made Mr. Johnson more skeptical and resulted in a bias about Randy's attitude and his adherence to standard operating procedure.

Summary

Here are a few principles that service professionals can employ to help ensure more empathy and less editorializing when working with customers:

- A person's concern for others implies the presence of empathy, which motivates people to put other's needs ahead of their own.
- A company owner's personal qualities culminate into a unique personality that creates a culture.

- Editorializing is what happens when a service professional says too much.

- For service professionals, the safety and well-being of customers during peak demand cycles is a primary concern.

- Going the extra mile for customers, while often an individual effort, is more important due to a pervasive attitude within a company—this is the culture.

- Empathy is the psychological software that allows people to care about those who need help.

- Technical expertise should be tempered with a dose of humility and empathy.

- Be wary of the Dr. Jeckle and Mr. Hyde service styles.

- Don't boost your own image by criticizing coworkers in the presence of a customer.

- The primary editorializing rule is "less is more," and it applies to what a customer hears during customer service situations.

- Editorializing occurs when a service professional shares too much information and this additional information detracts from customer satisfaction.

- Editorializing can diminish a customer's perception of your service company.

- In all customer interactions, what matters most is not what a service professional says, but what a customer hears.

Words and Weapons

5

The Power of Common Courtesy

Our mothers in many cases probably gave us our first and best customer service lesson when they taught us to say "May I," "Please," and "Thank you." These polite words still work wonders toward building rapport with customers. The correct word along with a positive tone of voice transforms a greeting into something special. How can something so simple be so profound? The answer is not always obvious. Innovative service companies have included the phrases "May I," "Please," and "Thank you" into their culture in an effort to differentiate themselves from their competition. Why? Because too many service companies do not inculcate this behavior as a core principle among their employees.

Customer expectations among homeowners regarding technical service professionals' communication skills are that manners and eloquence are secondary. The image that customers have regarding what a technical worker should look and sound like are based on personal experience or hearsay. These biases sometimes run deep. Yet this bias is an opportunity for service companies to gain a competitive edge and strategic advantage.

A profound opportunity to differentiate your service company can be achieved in how a service professional answers his phone when a customer returns a voice-mail message. A customer who calls a service professional

"Using the phrases 'May I,' 'Please,' and 'Thank you' has a dramatic and positive effect on every customer encounter."

back has no expectation beyond the information to be shared. However, when a service professional begins each returned phone message with the phrase, "Thank you for returning my call," then the customer hears the something special. What the customer hears is almost always more important than what a service professional says.

When asking a customer for their address, a service professional should ask, "May I please have the street address?" because this provides further courteous differentiation.

Using the phrases "May I," "Please," and "Thank you" has a dramatic effect on a customer encounter. The infusion of politeness results in two key benefits. The first thing it does is to surprise and delight customers, since the service professional's communication style is unexpected. The customer's surprise is usually audible. This also helps establish a calm tone and demeanor during the call. The second benefit is the service professional's halo effect, which results in greater customer flexibility. Both benefits simplify subsequent work.

Companies that are adamant about having their employees grasp the importance of saying, "May I," "Please," and "Thank you" often experience an uptick in their business. It never ceases to amaze me how surprised clients are with results that are based on simple things that make common sense.

Blame Implication and the "Do It Yourself" Customers

When service professionals encounter the "Do It Yourself" customer, they should minimize blame implication. The term *blame implication* can be defined as "a verbal phrase or nonverbal mannerism that implies the other person is at fault." Words can be weapons.

In this case it means that once a service professional has ascertained that the customer screwed things up, it's best not to make things worse by implying blame. Customers feel bad enough when their good intentions spiral downward to defeat. A talented service professional knows how to leverage these events to his advantage by making the customer feel better and thereby enhancing the rapport for future business opportunities.

The important role that emotions play in these customer interactions must be understood. A customer who knows he screwed things up and then gets a reprieve experiences emotional relief. Being spared the blame and embarrassment stirs positive emotions of appreciation for the service technician's tact, and this can result in greater rapport. This rapport can be an opportunity for a technician to forge a stronger business relationship and to ask the customer for referrals.

Another way to make a customer feel better is to pay them a compliment. The compliment doesn't have to be anything grandiose. One example of a compliment might be, "Unfortunately some customers don't contact me until the unit is damaged beyond repair. Contacting me was definitely the wise thing to do." Regarding compliments, only two types of people

FIGURE 5-1

The "Do It Yourself" Customer

His air-conditioning unit stopped working suddenly. Seeking to save money, Ed Clark decided to fix it himself.

Assuming the problem originated in the power supply, he delved into the unit and detached and reattached wiring harnesses in the hope of restoring electrical continuity.

When that failed, he probed, pressed, and bent sensitive components until he finally gave up and called a technician.

When the technician arrived at Ed's home and diagnosed the unit, he found evidence of tampering along with minor damage.

When the technician inquired about the damage, Ed played coy and spoke vaguely about what had happened. It took about 15 minutes of qualification before Ed admitted that he had attempted to fix the unit himself.

Rather than reprimand Ed, the technician thought it would be best to minimize blame since Ed probably felt bad enough.

Therefore, the technician continued his diagnostics, found the problem, and completed the repair.

And thus begins a long-term business relationship.

like to hear a compliment—men and women. A compliment needn't be too gushy or pretentious. Be subtle, concise, and straightforward and then get on with the business of adding value.

Service professionals who master the art of minimizing blame implication stand a better chance of explaining the benefits of scheduled maintenance by a professional to enhance their system's performance and to avert future equipment failure. Service professionals should not use too much technical jargon when explaining benefits.

Service professionals can sometimes convey too much scientific or mechanical jargon that is a manifestation of their expertise. There is no doubt that a good service technician knows his stuff. However, balancing expertise and empathy, while explaining the benefits, is a winning formula. Customers who hear and sense a service professional's empathy are more likely to invest in a long-term business relationship. The customer won't care how much you know until they know how much you care.

The biggest benefit of conveying empathy and not implying blame is that it keeps a customer from becoming defensive. When customers become defensive, this triggers an internal stress response that diminishes their ability to cooperate. Stress is often referred to as the fight-or-flight response. The fight-or-flight response has a specific meaning. Our ancient ancestors experienced stress when they fought with (fight) or ran away from (flight) predators. Stress is designed to enable a person to overcome a potential threat. In today's world, the predators have been replaced with modern, self-imposed stress because of the way we react to excessive workloads, time constraints, multitasking, upset customers, unhappy coworkers, and health concerns.

Service professionals experience the fight-or-flight response in less physical terms. Instead of running away from or fighting with a threat, some service professionals attempt psychological attempts of the same. Psychological "fight" occurs when a service professional gets so annoyed that he uses a snippy tone of voice along with nonverbal gestures that clearly express confrontation and displeasure. This behavior is usually retaliation for another person's bad behavior. Service professionals should practice restraint and not retaliation when working with difficult people.

Remember when your mother said, "It's not *what* you're saying, it's *how* you're saying it," in responding to your tone of voice? The same rule applies when service professionals interact with customers and coworkers.

The easiest way to practice restraint is developing a new habit, which includes a 2-second pause before making any verbal or emotional response. A 2-second pause gives a service professional time to think—to be constructive and rational in the midst of chaos. Conversely, an abrupt and instantaneous reaction will most likely result in the escalation of an adverse event.

During a 2-second pause, its best to take a slow and silent deep breath. The physical act of breathing deeply facilitates resourceful and innovative thinking, and it helps reduce stress and improve outcomes. The stress response usually results in fast, shallow breaths that diminish the oxygenation

"*Not implying blame when a customer screws up their equipment allows the customer to remain calm and not become defensive.*"

of the brain. Less oxygen means less constructive thought. Respiration is something a person can control immediately, and it must become a habit during adverse events. Conscious deep breathing relaxes muscles; and it causes the endocrine system to release endorphins, which are the body's natural painkillers. Needless to say, a slow, deep breath during a pause has numerous physiological benefits.

The pause and slow, deep breath should be followed by rational thinking. Rational thinking implies logical, reality-based thought rather than emotion-based thought. There is an element of calculation and planning involved in rational thought. It's an objective process and an analytic approach to solving problems. Thinking rationally after a pause and a slow, deep breath is bound to deliver more positive results to any problem.

Psychological flight occurs when a service professional's empathy and genuine concern flee, resulting in emotional detachment and apathy. Physically, a service person is still present; but emotionally, he has withdrawn. In this circumstance, the attitude of a service professional is "It's all about me" rather than "It's all about the customer," and this attitude diminishes a mutually beneficial outcome. Therefore, it's vital to believe that the customer is always right and behave accordingly.

Words, Health, and Lifestyle

Enduring the rigors of hostile work environments, such as when an HVAC technician works in an attic on a hot summer day, can also affect a service professional's demeanor. An HVAC technician working on an attic blower can experience temperatures greater than 140°F and sometimes as high as 170°F. Extreme temperatures can result in dehydration, which can affect basic physical and mental functions.

None of our body's cells or organs can work properly in the absence of water. Dehydration occurs when the amount of water leaving the body is greater than the amount being taken in. We lose water routinely when we sweat to cool the body and as humidified air leaves the body through exhaling. In extremely hot environments, water loss increases exponentially. The reduced fluid in the body's blood vessels affects cardiac output, so the blood vessels constrict in an effort to maintain proper blood pressure and deliver blood to vital organs. Eventually the reduction in blood affects bodily functions; and with severe dehydration, confusion and weakness occur as the brain and other body organs receive less blood.

Technicians working in extremely hot environments should alert their dispatcher or office reps and agree on follow-up calls every 15 to 20 minutes, so that a dispatcher regularly phones the technician to make sure he is still alert and functional. There are numerous accounts of technicians who have fainted and lost consciousness while working in extremely hot environments. An ounce of prevention with 15- to 20-minute follow-up calls can avert such emergencies.

Diet dramatically affects a service professional's overall performance and demeanor. When service professionals are out in the field, it's best to make smart food and snack choices. This recommendation is based on the impact that unhealthy eating can have on a service professional's ability to stay focused and perform well.

Limiting sugar intake during the course of your workday is a good start. Too much sugar may provide a quick energy boost, but the effect soon wears off and leaves service professionals more tired, irritable, anxious or depressed. None of these results coincide with the optimal behaviors needed to stay sharp, alert, and courteous.

The U.S. Food and Drug Administration (FDA) Dietary Guidelines recommend choosing foods and beverages with little added sugars. These are not the naturally occurring sugars in fruits, but rather those added in the processing of soft drinks, candy, cake, cookies, pies, and fruit drinks.

Sugars such as fructose, sucrose, honey, and corn syrup are metabolized in the same way. Likewise, the sugar found in a banana and that in a can of soda are both digested similarly, but the banana contains added benefits. The main difference is that most soda and candy are nutritionally empty, whereas the banana contains fiber, vitamins, minerals, and antioxidants. Bananas are handy and easy to carry (about the same size as a candy bar), and they have tremendous health benefits. The FDA advises people to eat more potassium-rich foods such as bananas, because potassium has a positive effect on blood pressure, brain functions, and stress.

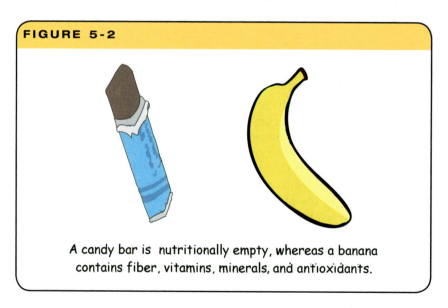

FIGURE 5-2

A candy bar is nutritionally empty, whereas a banana contains fiber, vitamins, minerals, and antioxidants.

A banana also contains an essential amino acid called tryptophan, which our body converts into serotonin. Serotonin helps regulate a person's feelings, and it enables a person to make rational decisions. Improved mental health is the required catalyst that can set your subsequent goal-setting activities in motion.

Service professionals who make wise food and snack choices enjoy improved lifestyle balance, greater mental clarity, and better health.

The Power of Words

Words have the power to soothe or exacerbate a situation. It all depends on their use and on the tone of voice that accompanies a service professional's speaking style and intent. You have probably heard a coworker use the expression, "He pushed my button," to describe a customer who intentionally antagonized a fellow employee. Yes, sometimes customers know exactly what words and mannerisms will "push your button."

During these adverse encounters, service professionals must remain calm and within the parameters of a strategy entitled "Contain, Qualify, and Correct." The following anecdote illustrates how Contain, Qualify, and Correct can be applied.

> *Looking rather exasperated, Rob Nicols, a plumbing and heating contractor, approached the sales counter at Action Plumbing Supply and dumped a hot surface igniter on the counter. When the counter salesman asked what was wrong, Rob yelled, "It's broken!"*
>
> *Speaking in a soft tone, the counter salesman told Rob how difficult it was when parts failed to work properly. Rob nodded with approval.*
>
> *The counter salesman complimented Rob for being smart enough to bring the hot surface igniter's original packing slip receipt, because this would help expedite a replacement. Rob liked being called smart. It made him feel good. Although Rob didn't know the counter salesman very well, Rob thought he was an OK guy.*
>
> *Feeling as though things were under control, the counter salesman asked Rob a qualifying question. The counter salesman knew that he needed a more detailed problem description than "It's broken!" Otherwise the manufacturer wouldn't reimburse his shop. So, the counter salesman asked, "What exactly is wrong with the hot surface igniter?"*
>
> *"There is a visible hairline crack—I checked it with a continuity tester," replied Rob. As he filled out the repair tag, the counter salesman agreed to replace the defective part immediately.*

The counter salesman in this story used the Contain, Qualify, Correct method of problem solving. Immediately realizing things were not under control, he knew the situation would need to be contained before he could qualify the problem and get an accurate problem description. Once the situation was contained and Rob was calm, the counter salesman asked a qualifying question. With a specific problem description, the counter salesman was able to correct the situation immediately.

As Rob departed from the supply house, he realized that the counter salesman had successfully resolved his problem with a minimal amount of effort. Then Rob thought about how he could apply this stable problem-solving method in his own contracting business. Rob understood that Contain, Qualify, and Correct enabled the counter salesman to get things done right the first time. It increased productivity, built rapport, minimized confusion, and enhanced customer satisfaction. *Contain* means you are getting the situation under control.

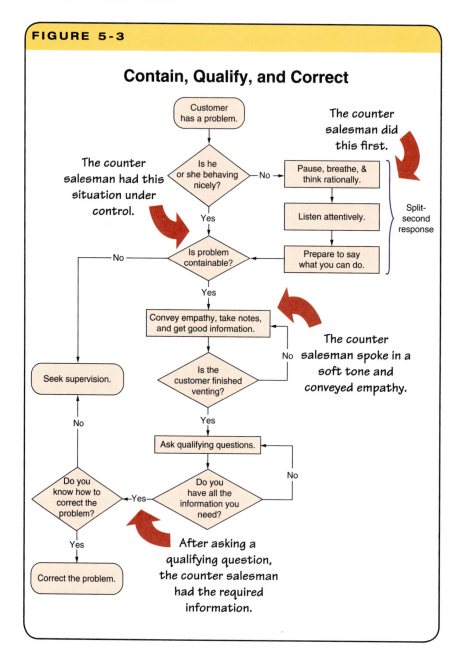

FIGURE 5-3

Contain, Qualify, and Correct

INSIGHT

A service professional's ability to ask the correct questions is a key tactic; therefore, qualifying the problem ensures a correct resolution. Once, while driving back to my office from a client meeting, I pulled into a rest stop on the Interstate Highway. I returned to my car after a brief respite, only to discover that the engine failed to start. I tried the ignition numerous times without success. There I was, stranded at a rest stop, quite conceivably at the mercy of emergency roadside service. Faced with the reality of the situation, I was reconciled with the thought of a very long delay.

My jumper cables were stored in the car, so I took matters into my own hands. A passerby was more than willing to help me. We connected our batteries using the jumper cables, after which I attempted to start my car. Unfortunately, we were unable to start the engine. With no other options, I phoned roadside assistance. After I described the known symptoms to a mechanic, he asked me two very important qualifying questions. First, "How old is your car?" and, second, "Do you still have your original factory-installed battery?"

In response, I informed him that my car was four years old and that it still had its original battery. With a clear sense of authority and certainty, he replied that in his judgment, the car battery was most likely dead. He informed me that his tow truck was out on a call, but that he would place my name next on his service list. I gave him my cell phone number and was instructed to wait until contacted by his driver.

My background as a service professional led me to conclude that this mechanic understood all about the importance of good qualification and a positive telephone vocal image. He knew that a person in my situation—that is, someone in need of help—would be put at ease by someone knowledge-able in his field of expertise. His telephone vocal image conveyed a sense of confidence and reassurance. I did not know who this individual was, what he looked like, or the type of environment he worked in; but based on his voice, I recognized that he was astute, alert, and methodical in his thinking and aware that each telephone call was a business opportunity.

I received a call on my cell phone a few minutes later from the tow truck driver, Ken, who politely introduced himself. He said that he was on his way to pick up a new battery for my car and would then proceed to meet me at the rest stop. Ken asked me about the make, model, color, and location of my car in the parking lot. He asked good qualifying questions that would enable him to do his job. In 10 short minutes a large tow truck arrived. Ken emerged from the cab and again introduced himself to me. He was quite young, very polite, and eager to be of service. His overall demeanor conveyed a sense of respect and appreciation that I was putting my trust in him. Ken did everything pos-sible to make me believe that my trust in him was well placed. Even though it was I who needed him and his services, Ken made me feel like it was he who needed me. What a concept! He made me, the customer, feel special.

Ken and his counterparts understood the importance of asking the correct ques-tions in an effort to get things done right the first time.

Rob continued to grasp the full impact of this methodology. The word *contain* simply means "to keep the situation within fixed limits so that the problem doesn't exacerbate." By keeping the situation under control, a contractor is able to do his job. Contractors need accurate answers to their qualifying questions. Imagine trying to get accurate answers from an emotionally upset customer. Emotionally upset customers usually do not provide accurate answers. Instead, they'll say things like "These things never work" or "It's always broken." Words such as *never, always, all,* and *every* are called absolute extremes. The absolute extremes are not very precise, and they don't provide the details a contractor needs to do his job.

Then Rob remembered that the counter salesman had paid him a soft compliment (saying he was smart) to defuse the tension. The counter salesman then proceeded to qualify in an attempt to get more details. This constructive interaction enabled him to keep the situation contained. Rob thought that paying customers a soft compliment when entering their home would work similarly. Perhaps complimenting a customer on an aspect of their home, their lawn, or something similar would be a great way to begin a new relationship. After all, business is all about establishing and keeping good relationships with customers.

The more Rob thought about it, the more this methodology made sense. Contractors who do not attempt to contain a situation run the risk of misdiagnosing problems and exacerbating situations. Not solving problems right the first time can result in rework, waste, and escalation of problems—all of which cost more time and money. If this happens, customers will blame the contractor for the mishap, thus further complicating the situation even though it was the customer's behavior that originally caused the confusion.

Perhaps the best reason to use the Contain, Qualify, and Correct method of problem solving is that it saves time. In any customer service situation, the faster a contractor can resolve a customer's problem, the better. There is a very strong correlation between response time and customer satisfaction.

The second reason to use the Contain, Qualify, and Correct method of problem solving is that it improves customer retention. Past studies have indicated that responding immediately and effectively to a complaint yields a 95 percent chance of customer retention. Not responding drops the retention rate to only 37 percent. To keep customers coming back, it's best to establish a stable methodology such as Contain, Qualify, and Correct.

Words and Voice Mail

Voice mail is merciless, unbiased, and highly efficient. It captures in great detail every nuance of a service professional's voice if he struggles through a cryptic, unintelligible, or rushed message. As a business tool, voice mail is nothing more than an input/output data device. You put your voice in, and the recipient takes your voice out. As with most business tools, the success

"Voice mail is merciless, unbiased, and highly efficient. It captures in great detail every nuance of a service professional's voice if he struggles through a cryptic, unintelligible, or rushed message."

or failure of a voice-mail message depends on cause and effect. If the message input is clear, articulate, and concise (the cause), then the recipient will likely hear and understand what to do next (the effect).

Most input/output business data devices experience the garbage in, garbage out (GIGO) phenomenon. In technical terms, GIGO occurs if invalid data is entered into a program; the resulting output also will be worthless. GIGO is usually a reference to the fact that computers, unlike humans, will unquestionably process the most illogical of input data, resulting in an unsound output. Therefore, if a service professional inserts garbage, then the recipient will likely receive garbage.

In voice-mail systems, garbage includes messages filled with one or more of the following:

- Background noise: Service professionals working at construction sites with excessive and audible machine noise should consider what the recipient will hear. Making the time to get inside the truck, before making phone calls, will result in a more coherent voice-mail message.

- Rapid pace of speech: Talking too fast makes it almost impossible for the person hearing the message to capture pertinent information. Some listeners have been known to replay the same message several times in an attempt to write down a telephone number. This clear waste of time frustrates the person listening to the message and also tarnishes the image of the fast-talker who left the message.

- Verbal junk: This usually manifests itself in the person who is uncertain of what to say and instead stumbles with sounds like "Uhhhm," "Duuuh," or "Hmmm." The voice-mail system captures their ramblings for public record as the caller attempts to pull together their ideas. Business professionals ought to be careful of what they say into voice mail since these messages are digital files that can be shared or distributed among others.

- Noncontiguous details: This phenomenon is a result of inadequate preparation—or the absence of any—before making a telephone call. The person leaving the voice-mail message skips from one detail to another with no regard to the sequential flow. Imagine then the person who listens to this message and tries to connect disjointed pieces of information.

Service professionals can also improve the quality of messages left by callers with a few innovative tactics. Voice-mail greetings can invite a quality message by asking callers to speak slowly, to state their telephone number twice, to leave only vital or important details, and to state the best time for a return call.

A service professional should acknowledge receipt of voice-mail messages with a return telephone call to the person who left the message. This is important even if the subject matter in the message will not be immediately handled, resolved, or dealt with. Acknowledgments go a long way toward keeping customers informed and at ease. Depending on the nature of your business, voice-mail greetings can also include information about your whereabouts; for example, whether you are in the office, out for part of the day, or on the road.

Voice mail is a powerful business tool in the hands of a competent person who understands the systemic implications of GIGO.

Summary

Here are a few principles that service professionals can employ to help ensure they use words thoughtfully when working with customers:

- Innovative service companies have included the phrases "May I," "Please," and "Thank you" into their culture in an effort to differentiate themselves from their competition.

- Service professionals who begin each returned phone message with the phrase, "Thank you for returning my call," enable the customer to hear something special.

- When asking a customer for their address, a service professional should ask, "May I please have the street address?" because this provides further courteous differentiation.

- Service professionals who encounter the "Do It Yourself" customer should minimize blame implication.

- A customer who knows he screwed things up and then gets a reprieve experiences emotional relief. Being spared the blame and embarrassment stirs positive emotions of appreciation for the service technician's tact, and this can result in greater rapport.

- Customers who hear and sense a service professional's empathy are more likely to invest in a long-term business relationship. The customer won't care how much you know until they know how much you care.

- The biggest benefit of conveying empathy and not implying blame is it that a customer does not become defensive. When customers become defensive, this triggers an internal stress response, which diminishes a customer's ability to cooperate.

- Service professionals experience the fight-or-flight response in less physical terms. Instead of running away from or fighting with a threat, service professionals make psychological attempts to do so with apathy and a snippy tone of voice.

- Service professionals should practice restraint and not retaliation when working with difficult people.

- Thinking rationally after taking a pause and a slow, deep breath is bound to deliver more positive results to any problem.

- Psychological "flight" occurs when a service professional's empathy and genuine concern flee, resulting in emotional detachment and apathy. Physically, a service person is still present; but emotionally, he has withdrawn.

- Extreme temperatures can result in dehydration, which can affect basic physical and mental function.

- Technicians working in extremely hot environments should alert their dispatcher or office reps and agree on follow-up calls every 15 to 20 minutes, so that a dispatcher regularly phones the technician to make sure he is still alert and functional.

- Diet dramatically affects a service professional's overall performance and demeanor. When service professionals are out in the field, it's best to make smart food and snack choices.

- The Contain, Qualify, and Correct method of problem solving saves time. In any customer service situation, the faster a contractor can resolve a customer's problem, the better.

- As with most business tools, the success or failure of a voice-mail message depends on cause and effect. If the message input is clear, articulate, and concise (the cause), then the recipient will likely hear and understand what to do next (the effect).

- Service professionals can also use a few innovative tactics to improve the quality of messages left by callers. Voice-mail greetings can invite a quality message by asking callers to speak slowly, to state their telephone number twice, to leave only vital or important details, and to state the best time for a return call.

Clutter and Clarity

6

Clarity, Cell Phones, and Cognitive Skills

The use of cell phones for phone conversations, text messaging, and e-mail has unfortunately become pervasive in our society. Text messaging and writing e-mail while behind the wheel are more dangerous than driving while under the influence of alcohol. A service professional's clarity and alertness requires mental focus. When service professionals focus their mental energy on one issue (cell phone), it can cause them to miss other important things (the road, pedestrians, other drivers, etc.).

A person's cognitive skills are negatively affected while using a cell phone; this effect is known as cognitive capture. Cognitive capture is a phenomenon in which the cell phone user is too focused on the instrument to observe the whole environment. For example, while driving, if the driver is focused on the text messaging and not on the road, he or she is suffering from cognitive capture. Service professionals expose their employer to financial liability when they use cell phones while driving.

Here are a few cell phone etiquette ideas to consider:

- When your cell phone rings while you are at the supply house counter buying parts and materials for the day's work, do not ignore the counter salesperson to answer your phone. His or her time is important.

If a call is important and you feel you must take it, be brief and let the caller know you'll call them back.

- Turn off your cell phone or set it to vibrate when you are in a customer setting that requires your full attention and a quiet atmosphere.

- If you are expecting an important call and you are going into a customer meeting, mention at the start that you are expecting a call that you will need to take.

- When you're eating out with a customer or coworker, texting on your cell phone below or above the table at a restaurant is considered rude. Keep your cell phone out of sight.

FIGURE 6-1

Cell Phones and Cognitive Capture

Cognitive capture is a phenomenon in which the cell phone user is too focused on the instrument to observe the whole environment.

Clutter and Clarity

Carl searched through the mound of papers piled up on the passenger seat of his truck. The accounting clerk at the office had phoned him with a simple question regarding a replacement part number. Carl could answer the question in 2 seconds, if only he was able to find the right paperwork. He handled every piece of paper in the pile twice, and he was still unable to find the correct work order.

Carl's rate of breathing increased, his heartbeat was racing, and he had a difficult time focusing on the task at hand. The more Carl searched, the more stressed out he became.

"Do you want to call me back?" asked the accounting clerk.

To which, Carl exhaled loudly and replied, "Uhm, duh," and other undecipherable mutterings.

"Just call me when you find it," said the accounting clerk as he hung up the phone.

The relationship between clutter and clarity is profound. Simply defined, clutter is a confused or disorderly condition in which a collection of items are not in their expected places. These items pile up in a conspicuous location, one item eclipsing the other, and the result is a mess. But wait, there's more! Clutter also has negative side effects that limit a person's ability to communicate with internal and external customers. Therefore, clutter is not just "the mess"—it also includes the damaging consequences resulting from the mess.

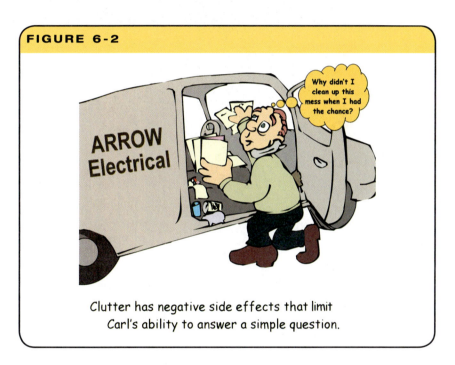

FIGURE 6-2

Clutter has negative side effects that limit
Carl's ability to answer a simple question.

Clutter does not happen all at once. It begins with a belief that "right now" is not the best time to put an item away. Procrastination then seeps into a person's behavior, even though he or she usually has the good intention of returning an item to its proper place sometime in the near future. The cause of clutter is the absence of a disciplined routine of putting items in their proper place immediately.

Clutter includes tangible items such as paperwork, spare parts, tools, debris, and other paraphernalia. The problem with clutter is that it diminishes a service professional's command over situations that are within his or her sphere of control. The presence of clutter is the root cause of numerous service company mistakes, miscommunications, and lost business.

In the preceding story, Carl spoke unintelligibly because his brain was dealing with two conflicting messages:

1. This is a simple question that I should be able to answer.

2. I am unable to answer this simple question.

Carl's brain got locked down in an endless cycle of "I should be able to" and "I am unable to," and the result was confusion. The stress response, which activates his body's defense mechanism, further diminishes his ability to engage in calm, creative, clear thinking. While clutter is the root cause, it is stress that renders Carl ineffective. No real threat exists except for the threat to Carl's ego. He knows that this embarrassing situation could have been averted had he taken the time to organize and put away his paperwork.

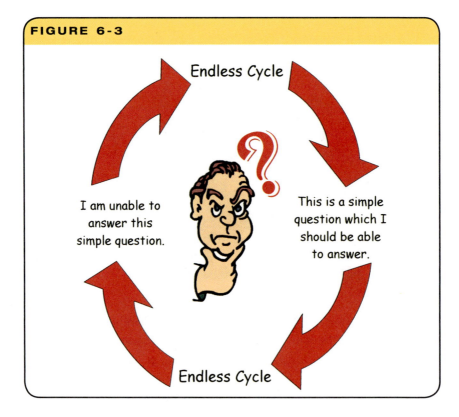

FIGURE 6-3

Endless Cycle

I am unable to answer this simple question.

This is a simple question which I should be able to answer.

Endless Cycle

Carl's thinking dwells on the internal questions such as, "Why did I procrastinate?" and "This is unlike me; how could I let this happen?" This self-destructive behavior involves dominant thoughts about defense and survival. This creates the internal battle that triggers his psychological fight-or-flight response (discussed in Chapter 5).

This aggressive and narrowly focused mind-set is in stark contrast to the calm perspective required to view the wide panorama of potential solutions. And this is what caused Carl's "uhm, duh" response when asked a simple question.

"The problem with clutter is that it diminishes a service professional's command over situations that are within his or her sphere of control."

The cumulative effect of procrastinating results in a reactive stance that is preventable. If Carl chose not to procrastinate but rather to organize his paperwork, he would benefit from a proactive stance, which would yield greater clarity, efficiency, and productivity. Yet when service professionals falter due to their own self-imposed hindrances, this reactive environment makes things worse for everyone.

Verbal Junk

Carl's unintelligible response is called verbal junk. As the term implies, verbal junk is what happens when a person conveys words that diminish clarity and substance. Carl could have remained calm and said, "I know it's here someplace, may I call you back after I find it?" This would have explained the delay and set future expectations for the accounting clerk. Instead, Carl descended into the downward spiral of verbal junk, the root cause of which is clutter.

The utterance of unintelligible responses such as "uhm, uh, or duh" is usually the manifestation of not knowing what to say. In this situation, a service professional feels compelled to say something; and in the absence of knowing what to say, the result is usually verbal junk. Feeling compelled to speak and not knowing what to say is a bleak condition to be in. It's best to remain calm and resist the pressure of feeling backed into a corner. Silence sounds better than awkward verbal junk, but silence might arouse the other person's concern about your responsiveness. Therefore, knowing what to say is vital.

Make time to think and remain constructive—this is the best strategy to minimize verbal junk. A calm investment of a few seconds will yield a much more positive response. When a customer hears either of these responses: "Good question; may I get back to you on that?" or "I can help you with that. While I'm looking, would you prefer to wait, or should I call you back?" then a more constructive interaction occurs.

The two questions in the previous paragraph are clear, concise, and comprehensive. What would happen if the first sentence looked like this: "Basically, that's a good question; may I sort of get back to you on that, you know?" The words *basically, sort of,* and *you know* don't add value or extra meaning to the sentence. These words are also verbal junk.

Unfortunately, verbal junk words such as *basically, sort of,* and *you know* are used pervasively, mostly because it has become a habit. When customers hear vague and verbose sentences, their perception of the speaker is diminished. Service professionals convey a more positive concept of their intelligence and expertise when they speak concisely and clearly.

A service professional's voice conveys five attributes to a customer:

- **Health:** A robust and healthy vocal tone is more attractive than a weak and nasal-sounding tone of voice. Good personal hygiene habits add to the positive perception of excellent health. Service professionals convey a more positive halo effect when these factors are considered.

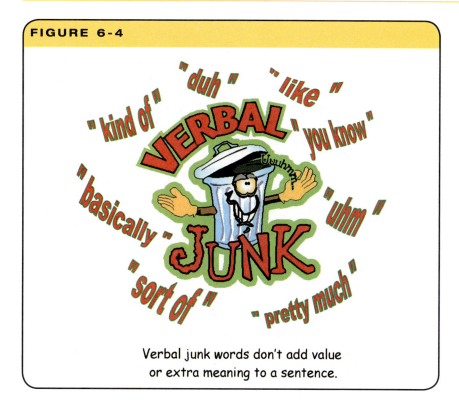

FIGURE 6-4

Verbal junk words don't add value
or extra meaning to a sentence.

- Intelligence: Investing more time in listening to customers conveys a service professional's intelligence and stature. Smart service professionals know it's best to ask questions and to qualify their understanding of an event rather than to make assumptions or to jump to conclusions.

- Education: Minimizing verbal junk and speaking in an articulate manner are ways for a service professional to enhance a customer's perception of their training and education. Service professionals who are selective about their statements tend to editorialize less and convey a more positive vocal image.

- Assertiveness: By starting with a positive attitude, service professionals can convey their tenacity and willingness to help the customer. Service professionals exhibit a more self-confident persona when they convey enthusiasm and eagerness in resolving customers' issues.

- Ability: Service professionals demonstrate their aptitude by maximizing their halo effect. Enhanced communication skills culminate in a positive first impression, during which customers are more likely to cooperate and listen to a service professional's suggestions and recommendations.

Keeping Customer Requests Constructive

Every service professional faces a similar dilemma on a daily basis when serving customers. Most often, this dilemma is the natural progression during a service visit. The encounter is moving ahead smoothly, and then the customer stops to ask a question about a product, service, or company policy. An unprepared service professional may feel anxiety build up, and he would prefer to say "Yes"; or "Certainly, this product can do that"; or "I can install it today." But due to peak activity or capacity issues, he will not utter any of those phrases today, because he must speak the truth. And the truthful answers are "No"; or "I am sorry, but this product does not do that"; or "I cannot install it today."

The quandary involved here is how to say "No." Saying "No" might make a service professional feel unpopular or like a killjoy. Conveying seemingly bad news to a customer might bring an end to all of the goodwill he has been creating while attempting to provide a service. Regardless of how a service professional might feel personally about having to say "No" to a customer, sometimes the answer must be "No." What separates the seasoned professionals from the amateurs in the service business are three distinct behaviors: (1) knowing when to say "No," (2) knowing why to say "No," and (3) knowing how to say "No."

The first key behavior involves a sense of timing and good listening skills. Many service professionals have an instinctual sense, when dealing with a customer, that they must respond in the negative. This "sixth sense" triggers a need to provide an immediate answer. The reply is communicated without any pause whatsoever on the heels of the customer inquiry. This is not an acceptable reaction, since timing is a vital communication skill. Instead, a service professional should allow himself a moment or two to formulate a qualifying question in order to ascertain the importance of the criteria. As an example, if a customer were to ask, "Can you install it today?" you might qualify your answer by saying, "If I could install it tomorrow, would that be OK?" Your question qualifies the criteria required to achieve customer satisfaction. Should the customer answer "Yes," then you have the flexibility to install the equipment on the following day. In this illustration, the technical answer to the question was "No." However, with a pause, and some time to formulate a rational response, and a bit of finesse, a much more creative answer surfaced.

There will be instances when the consequences of doing what a customer requests will outweigh any benefits. One such case might arise when a service professional risks the stability of his company's system in order to satisfy the request of a specific customer. In that case, it is important to be sure the customer understands why you cannot carry out the request.

The concept of knowing how to say "No" begins with an adherence to the fundamental principle of saying what you *can* do rather than what you *cannot* do. When service professionals convey what they can do, it keeps the proverbial door open so that the dialogue with the customer continues their business relationship. However, when service professionals resort to what they cannot do, it threatens to limit future dialogue as well as any future business.

Some customers have very high expectations, and they will make excessive demands. It is vital that service professionals maintain a constructive demeanor during these encounters to ensure customer satisfaction.

Personal Congruency and Clarity

This book reinforces why congruency is such an important factor in delivering world-class service. On a personal level, congruency also plays an important role. When a service professional's behavior is congruent with his or her belief system and personal constitution, then clarity usually ensues due to the absence of guilt and embarrassment. Carl's internal conflict, described earlier, is what happens when a service professional's behavior is incongruent with his true self.

A person's belief system and personal constitution about work usually begins at home. Parents help inculcate the disciplines and good habits that translate into professional workplace behavior. Remember when your parents urged you to clean your room or put your toys away?

The most meaningful lessons I learned about work, life, and serving others came from working with my father. He was born in 1912 and managed to achieve only a fifth-grade education. Students who were unable to maintain specific academic standards in those days discontinued their education after the fifth grade in order to enter the workforce. I suppose my father just did not learn the way other children did. In subsequent years, he moved from one manual-labor job to the next. This was no problem for my father, because he was stronger than most.

My father eventually began working with construction crews, where he developed a skill for working with concrete. Even without a formal education, he was able to teach himself the use of a slide rule so that he could calculate the amount of materials required to complete a project. Although the mathematical computations were an exact science, my father possessed something intrinsic. His skills included a visceral understanding of the texture, density, and granularity of concrete.

Many a Saturday of my boyhood was spent accompanying my father on one of his smaller concrete projects such as driveways, patios, and walkways. We would rise early to load a large concrete mixing bin on top of his 1959 Ford Ranch Wagon. After gathering the necessary shovels, hoes, trowels, levels, and tool boxes, we drove to the worksite.

My father taught me the importance of knowing how to select and use the correct tool for the appropriate job. I also learned the importance of properly storing the tools so that we would be able to find them when needed. I learned that being organized is a key to success, whether I was a laborer or a consultant. The discipline of minimizing clutter and not procrastinating about putting the tools away was a lesson that stuck with me. It became part of my belief system and personal constitution. Occasionally random lapses of the "putting the tools away" discipline occur, and thankfully I catch myself and revert to my initial instruction.

A service professional's concept of value, personal responsibility, and integrity is a result of parental instruction. My understanding of truth and integrity began at home. So deep was this understanding that I rarely had to think twice about the difference between what is right and wrong.

"I learned that being organized is a key to success, whether I was a laborer or a consultant. The discipline of minimizing clutter and not procrastinating about putting the tools away was a lesson that stuck with me."

Back in the 1970s, one of my most memorable part-time jobs involved selling industrial cleaners and degreasers over the phone to contractors. It was a telemarketing firm, and each outbound rep was given a stack of index cards to call during the shift (this was before computers and cell phones). The index-card stacks were a key to the outbound calling infrastructure. Each card included a prospect's contact information. It was hard work only because the companies I needed to phone had already been called by a previous part-time rep. This was a setup for lots of rejection.

From the experience, I learned the importance of a good script along with an enthusiastic tone of voice. On this job, I was also trained to use the "steak knife pitch"—buyers could earn a free set of five stainless steel steak knives with each order. It was a corny sales pitch, but it actually worked. Paul, a guy who sat next to me, was promoted to making the "camera pitch," which enabled buyers to earn a free Polaroid camera. During each work shift, our ears were filled with a cacophony of cold calls and sales pitches, all going on at once.

My sales were sporadic; but with each sale, the index card for that purchaser went into a special stack, categorized by sales rep. The special stack was reserved for the "Death Pitch." I didn't know what the "Death Pitch" was, and no one would tell me. I anticipated the answer to what remained a big secret.

One day, Paul announced he was quitting because he had found a better job. His special card stack of purchasers was handed to a supervisor. The supervisor's job was to deliver the "Death Pitch." Finally, I would learn about the "Death Pitch." So as the supervisor called the first customer from Paul's stack, I sat idly by and listened. This is what I heard:

> Hello Mr. Smith, this is Gary from Regional Chemical Labs. I am calling with bad news. It's about Paul, our sales rep who has handled your account. As you may know, Paul was an avid sportsman and an airplane pilot. Last week, while he and his wife were flying in Paul's plane, a major malfunction caused the plane to crash. Perhaps the only blessing about this fatality is that Paul's two children were not in the plane. Our company is setting up a trust fund for the children by selling Paul's entire warehouse inventory. Do you think you could help us liquidate the inventory with an additional order that will benefit Paul's children?

Needless to say, that was my last day on the job. I quit that afternoon. The blatant dishonesty was incongruent to my parents' instruction and my personal constitution.

Years later, as a service manager I helped create a culture of honesty by training my staff to practice a very basic principle: it's better to disappoint a customer with the truth than to satisfy a customer with a lie. This principle does not imply that service professionals should upset customers with disappointing news. Instead, it implies that how a service professional pauses, thinks rationally, and conveys the truth in a constructive manner makes all the difference.

Summary

Here are a few principles that service professionals can employ to help minimize clutter and increase clarity when working with customers:

- Clutter has a negative side effect that affects a person's ability to communicate.

- The cause of clutter is often the absence of a disciplined routine of putting items in their proper place immediately.

- Clutter diminishes a service professional's command over situations that are within his or her sphere of control.

- While clutter is the root cause, it is stress that renders a service professional ineffective.

- Stress heightens a service professional's metabolism, and this aggressive and narrowly focused mind-set is in stark contrast to the calm perspective required to view the wide panorama of potential solutions.

- The cumulative effect of procrastinating results in a reactive rather than a proactive stance.

- Verbal junk is what happens when a person conveys words that diminish clarity and substance.

- Unintelligible responses such as "uhm, uh, or duh" are usually the manifestation of not knowing what to say. Making time to think and remain constructive is the best strategy to minimize verbal junk.

- A service professional's voice conveys five attributes to a customer: health, intellect, education, assertiveness, and ability.

- When seasoned professionals are unable to fulfill a customer's request, these three tactics will keep the dialogue constructive: (1) knowing when to say "No," (2) knowing why to say "No," and (3) knowing how to say "No."

- The concept of knowing how to say "No" begins with an adherence to the fundamental principle of saying what you *can* do rather than what you *cannot* do.

- Cognitive capture is a phenomenon in which the cell phone user is too focused on the instrument to observe the whole environment.

- When service professionals focus their mental energy on one issue (cell phone), it can cause them to miss other important things (the road, pedestrians, other drivers, etc.).

- When a service professional's behavior is congruent to his or her belief system and personal constitution, then clarity usually ensues due to the absence of guilt and embarrassment.

- A person's belief system and personal approach to work usually begin at home. Parents help inculcate the disciplines and good habits that translate into professional workplace behavior.

- It's better to disappoint a customer with the truth than to satisfy a customer with a lie.

Listening
and Learning

7

Did You Hear That?

"What style of lighting fixtures did you envision in your family room, Mrs. Stanton?" Aaron asked. This question reminded Aaron that he had forgotten to order new fixtures and a 200-amp circuit panel for another customer. The order needed to be phoned in today to meet a deadline. When Mrs. Stanton answered Aaron's question, he wrote himself a reminder about the other customer's order.

Looking at Mrs. Stanton, Aaron realized she had conveyed her lighting fixture preference while his mind was preoccupied. He had no idea what Mrs. Stanton had said because he wasn't listening.

He made eye contact and demonstrated good nonverbal skills, but his mind had drifted into la-la land and now he was in a tight spot. She had mentioned recessed high-hat fixtures earlier, but was this her answer? Aaron didn't know.

In a split second, Aaron fearlessly mustered the necessary courage and asked, "I'm sorry, Mrs. Stanton, what was your lighting fixture preference again?"

Aaron focused this time. "I think the track lighting will work best," she answered. He wrote down Mrs. Stanton's preference and asked a few more qualifying questions.

FIGURE 7-1

Aaron's mind-drift caused him to lose valuable information—he wasn't listening.

Aaron's mind-drift experience can happen to almost anyone. Mind-drifting (or mental wandering, wandering mind, daydreaming, la-la land, etc.) is a cognitive phenomenon in the brain wherein one's attention becomes distracted from the task at hand and strays into unrelated thoughts. Busy service professionals engaged in multitasking activities will experience numerous daily mind-drift moments.

The average person experiences mind-drift about 30 percent of their waking day. Much of this includes the typical daydreaming that might occur while driving a vehicle or taking a walk through a park. Mind-drift becomes a problem when it interferes with a service professional's listening skills. Five or 10 seconds of mind-drift, while a customer is speaking, can result in a loss of valuable information for a service professional. The best remedy is fearless and courageous listening behavior to ensure the message is heard, qualified, and understood.

When people think about listening, they assume it is similar to hearing. This is a precarious misconception because it leads people to believe listening is passive. Hearing a message is a passive exercise; but listening to a message requires mental energy, and this makes listening skills more active. To fully understand a spoken message, a listener must hear, qualify, and understand what is being spoken.

Service professionals may get bombarded with multiple messages simultaneously, and this makes it almost impossible to listen effectively. In the midst of a multitasking and busy environment, service professionals must be fearless qualifiers. Most adults are cowardly qualifiers, because they lack the courage to go the extra step to fully understand what is being spoken.

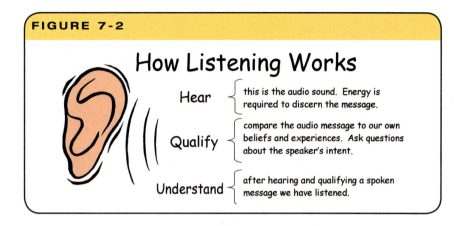

FIGURE 7-2

How Listening Works

Hear — this is the audio sound. Energy is required to discern the message.

Qualify — compare the audio message to our own beliefs and experiences. Ask questions about the speaker's intent.

Understand — after hearing and qualifying a spoken message we have listened.

Here are some of the most common cowardly listening behaviors:

- Too overwhelmed with the technical jargon being conveyed by someone else—assume the message is understood, even though confusion prevails.

- Too embarrassed to admit mind-drift—continue the hoax and hope for the best.

- Too timid to ask a person with a heavy foreign accent to repeat themselves—fail to qualify.

In Chapter 3, this book delves into a person's bias as the tendency to make assumptions based on our own filter, which is a culmination of beliefs, experiences, and lessons learned. It is this personal filter that adds complexity to the listening process.

Because everyone filters words through their own personal bias, no two people have exactly the same meaning for words or expressions. For example, the dictionary contains thousands of words; however, the average adult uses about 500 of these words most often, and each word can have multiple meanings. Biases will cause different people to interpret words differently—it is a subjective exercise. A word is merely a representation of something based on a person's beliefs, experiences, and lessons learned—namely, their bias.

Therefore, words can be ambiguous, as these examples demonstrate:

- A convenience store displays a sign for hungry customers. The sign reads, "Eat here and get gas." The convenience store owner's intention is to sell gasoline, but some customers mistake the sign as a gastronomical warning. The word "gas" is ambiguous.

- An electrical parts supply house sign reads, "Customers who think our counter salesman is uncivil ought to see the manager." Here the word "see" is ambiguous.

Mutual understanding improves when we engage more of our sensory information. Taking notes while listening utilizes two senses: listening and writing. Our ability to retain information increases when we take

"A person's ability to hear, qualify, and understand increases exponentially when they do three things: listen, take notes, and echo what was spoken."

notes. Take it a step further and repeat the information back to the customer (this is called the echo technique), and you engage more sensory information. A person's ability to hear, qualify, and understand increases exponentially when they do three things: listen, take notes, and echo what was spoken.

Short Attention Span

The average attention span has diminished due to the rate of interruption that service professionals must endure. While interruptions come in many forms, the most pervasive interruptions are portable communication devices. Incoming phone calls, text messages, and e-mails can be ongoing. Each interruption requires mental energy and focused attention about whether to reply immediately or defer a response until later. For field service professionals, there is ongoing contact from a dispatcher regarding scheduling changes and updates.

Turning off the ringer on a portable communications device does not always help matters. Some service professionals will make matters worse by placing their portable communications device on vibrate mode before a customer meeting. While vibrator mode is inaudible, it still becomes a listening distraction. Have you ever noticed the lack of concentration as someone ponders their uncertainty about whether they should glance at their portable communication device as it vibrates on their hip? This is a silly predicament for a service professional to be in. Sometimes it's best to turn off a portable communications device. Service professionals must discern when their listening skills need to be sharpest and then take the necessary precaution.

If you experience mind-drift while listening to a customer, it's best to assess whether you missed anything important and then act courageously. Say the following, "I want to make sure I heard that last part correctly; may I please hear that again?" Asking questions is better than fixing mistakes due to miscommunication.

Biases and Listening

Listening skills are diminished when a service professional's bias results in a preconceived notion about others. Ask a female trade technician if she has ever been the victim of gender bias, and she will answer in the affirmative. Biases can be about gender, age, ethnicity, attire, where a person lives, the type of car a person drives, the way a person speaks, and so on. A man with preconceived notions about female trade technicians may not listen as intently, because women do not fit within his expectation of a trade technician.

The same biases apply to male technicians who look too young for their age. Customers may assume that a young-looking technician is less experienced and will deliver inferior service.

It is important for service professionals to stay calm when others make incorrect assumptions based on bias. Service professionals must always practice restraint and not become defensive when these bias situations arise. Service professionals who get upset will make things worse with self-destructive behaviors that diminish rapport and minimize good listening skills.

In all customer encounters, service professionals invite the type of behavior that they convey toward customers. Adverse behavior only begets more adverse behavior—it's best not to get there in the first place.

The best listeners are courageous, and they trust that others will appreciate qualification. Business rapport is enhanced when a customer senses that you care enough to ask questions and verify the details. Unfortunately, some adults believe that asking too many questions might be perceived as a lack of intelligence. This is nonsense. If you need information, ask for it.

Would You Rather Be Right or Happy?

Some people stop listening when they assume the other person is wrong. This assumption diminishes a person's ability to understand another's intent and feelings. The need to be right is usually driven by a personal agenda, the ego, or lack of humility. While service professionals don't have to agree with others, it is important to acknowledge the way others feel.

When service professionals assume that the other person is wrong, they are more inclined to interrupt in an effort to correct the mistake. It's best to allow others to finish their complete thought and then ask a therapeutic question designed to open up a constructive dialogue. An example of a therapeutic question is, "Can you help me understand why you feel that way?"

Help Me to Help You

Most people, when asked to help another person, will agree to assist. Most of us have a desire to please and will muster the required time, energy, and cooperation to help another human. This social dynamic brings people together and builds mutual rapport. Throughout my management career, I have based much of my policy and personnel assumptions on the premise that people are good, generous, and trustworthy. Therefore, I believe that people will come to the aid of others when a need arises.

The HVAC industry has seasonal peaks and valleys relevant to business activity, and companies allocate their installation, repair, and maintenance work accordingly. This means that when thermal extremes arise, such as in the winter and summer, most HVAC activity peaks with a plethora of repair and installation assignments. Then, during the milder seasons, HVAC companies perform mostly maintenance work to ensure that their customer's equipment is optimized for when the extreme weather does arrive. The HVAC dispatchers and service reps schedule each field technician's time by establishing priorities based on first-in, first-out (FIFO), customer needs, and equipment failure.

Every now and then, during the milder seasons of spring and fall, the weather changes abruptly; and this begins a chain reaction of unexpected events. This chain reaction usually begins on a day in mid-November, when field technicians are prescheduled to perform routine maintenance work. These maintenance assignments were scheduled weeks in advance with the assumption that the average temperature would be 40° to 50°F. However, if Mother Nature throws the preverbal monkey wrench into the works and the temperature dips down into the 20° to 30°F range, then field technician's assignments need to be rescheduled.

The rescheduling is done by the dispatchers and service reps. These calls are awkward because customers can perceive the message as saying, "Someone else's needs are more important than yours, so therefore we must reschedule."

The remedy that I suggest involves a simple and profound change: ask a therapeutic question rather than make a statement. The therapeutic question I suggest begins with two of the most important words in the customer service lexicon: *help me.* The script goes like this: "Hello, Mrs. Smith, this is Steve from Action HVAC. Help me understand whether you have any flexibility regarding today's maintenance visit. Might we be able to reschedule?"

The script just presented changes the demeanor of each outgoing phone call to a constructive dialogue in which the customer's input is invited. Some customers lack flexibility due to work or family constraints; but more important, other customers will be able to cooperate because of the outside weather. Customers who are able to cooperate should be sent an unexpected token of appreciation. A handwritten thank-you note is an inexpensive, yet profound gesture that customers will remember. To make an even stronger impression, include a coupon or other value-added premium in the envelope. Small, inexpensive, and yet profound gestures forge a bond between service companies and customers. Naturally, this bond yields greater customer retention and more positive word-of-mouth referrals.

Another very powerful force is also at work—the law of reciprocation. Numerous sociology studies prove that we feel obliged to repay someone when they do something nice for us. The law of reciprocation possesses awesome strength, often producing a yes response to a request that, except for an existing feeling of indebtedness, would surely have been refused.

The law of reciprocation is the most powerful form of persuasion and influence.

The Law of Reciprocation

One Ohio HVAC company inserts a one-dollar lottery ticket in thank-you notes to customers who are flexible about scheduling changes. The lottery ticket is perceived as an accommodation, and customers remember the courteous gesture. The company owner told me that after phoning a residential customer about rescheduling service mainte-nance, the customer asked, "Does this mean I get another lottery ticket?" Apparently, the lottery ticket is memorable and leaves a positive and lasting impression.

Customer retention is easier to achieve when service professionals exceed a customer's expectations using the law of reciprocation.

A technician who arrives at a customer's home and sees a newspaper in the driveway should bring the newspaper to the front door. A customer's expectations are exceeded when a service professional's courteous greeting is accompanied by the delivery of today's newspaper. The subtle indebtedness a customer feels might be conveyed in the form of greater flexibility and more cooperation later on.

The law of reciprocation is influential, and it works. Service professionals should leverage this powerful strategy with genuine and sincere motives. When your customers sense your good intentions they will, in the name of reciprocity, give back more than they have received in the form of future business.

Mutual Understanding and the Three F's

The answer to a therapeutic question can begin a constructive dialogue using the three-*F* method of mutual understanding. The three-*F* method involves conveying the words *feel*, *felt*, and *found* in methodical and sequential sentences.

An example of the three-*F* method is:

- I understand how you *feel*.
- Others have *felt* similarly.
- What they *found* is that a preventive maintenance agreement saves money over the long term.

Asking Questions

One of the most therapeutic listening enhancement techniques involves using the words *help me* as a prefix to a question.

Some examples look like this:

- Can you *help me* understand what went wrong with the hot water heater?
- Can you *help me* understand if your schedule is flexible today?

INSIGHT

Telephone Therapy

When service professionals listen to customers, they are providing a form of therapy. This can happen in face-to-face situations or over the phone. Listening over the phone is more difficult due to the absence of visual clues, such as facial expressions and hand gestures. Therefore, listening over the phone requires a higher level of concentration and focus. In addition, the visual distractions that can arise when speaking over the phone can minimize listening skills.

A good psychotherapist invests more time listening than speaking. When patients feel like they are being heard and understood, they feel better. Likewise, customers feel better if they are being heard—even over the phone.

The *Journal of the American Medical Association* published a study on the benefits of psychotherapy using the telephone. A psychiatrist at the Group Health Cooperative in Seattle reported that 80 percent of their patients who received telephone therapy along with antidepressants indicated that their depression was "much improved" 6 months later, compared to 55 percent of those who received medication alone. The inception of the telephone therapy study was in response to the increasing number of patients who failed to maintain their in-person counseling sessions long enough to detect any benefits. One out of every four patients attending in-person psychotherapy drop out after just one session; fully half would cease treatment altogether by the fourth session. A psychiatrist responded to this trend by contacting patients by telephone to find out whether that method made it easier for them to continue with their treatment sessions. It did. The telephone can be as therapeutic as a face-to-face visit.

The words *help me* are therapeutic because they invite the other person into a more constructive interaction. This invitation, while not overtly stated, enables you to make the other person part of the solution. Simply, you are inviting the other person to help you.

A good therapist knows how to ask questions to set up subsequent listening. Open-ended questions can be used to expand a discussion in the interest of receiving more details. Close-ended questions prompt specifics in one- or two-word answers.

Examples of each are as follows:

- Open-ended: What symptoms did the air conditioner exhibit?

- Close-ended: Is the air conditioner working properly?

The Mutual Benefit

Improved listening skills enable service professionals to establish trust and better customer rapport. Enhanced rapport can yield many mutual benefits. As the term implies, *mutual benefits* means that both the customer and the service company win. The enhanced relationship is most important when a service professional makes a suggestion about additional services and equipment. Customers are more likely to invest in additional services if trust and rapport have been established.

While most service professionals are paid to repair, install, and maintain the equipment in a residential customer's home, the opportunity to upsell, providing a mutual benefit, exists. The definition of *upselling* is "Making a suggestion to an already receptive buyer to enhance the value of his or her purchase." Service professionals need not be too assertive when making an upselling suggestion. An upselling suggestion is a polite nudge along with a smile, genuine concern, and a little persuasion. It is not an aggressive tactic.

Congruency and upselling suggestions go hand in hand. Service professionals who suggest additional services and equipment that are congruent achieve more upselling success. This is called the natural pairings approach to upselling suggestions. Services and equipment that naturally go together have more congruency, and therefore customers are less likely to become confused or skeptical.

"Do you want fries with that?" Millions of additional dollars have been made with that simple question. The reason the upselling question works so well is congruency. Fries and burgers are a natural pairing.

A service professional who repairs HVAC equipment may suggest a natural pairing such as a service maintenance agreement, a humidifier, or indoor air-quality equipment. All three suggestions are congruent to HVAC equipment, and both the customer and the service company win in the presence of a service professional who can convey the benefits.

Following Up

What do the following situations have in common?

- A mother waits anxiously by the front door for her teenage daughter, who was supposed to be home 2 hours ago. The daughter hasn't called, and the mother worries.

- When all of the executive finance managers are called into an emergency meeting, the receptionist begins to fret about the company's future and her job.

- A man takes the day off from work so he can be home for a new cable TV installation. It's almost 5 p.m., and the homeowner wonders if the cable company forgot about him.

In these three scenarios, there is an absence of information. The mother, the receptionist, and the homeowner reacted to their experiences by expecting the worst. Most people would react similarly. In the absence of information, speculation fills the vacuum. And all too often, the most common human response is negative. Therefore, following up with a customer fills the void with up-to-date information so customers will not assume the worst.

Why follow up with customers? Because following up with a customer surfaces more information, and this has financial and relationship value. It is information that allows customers to make more choices or to obtain greater satisfaction than they would from choices made in the absence of information.

Following up with customers about significant and moderate events helps minimize misunderstandings and maximize rapport. In addition, customers who are well informed are less likely to be skeptical and second-guess your recommendations and suggestions.

Strategic follow-up, as part of a standard operating procedure, is a key differentiator among service professionals. Those who practice discipline and consistent follow-up enjoy improved and more profitable customer relationships.

If following up is so important, then why don't more service professionals do it? The answer lies in the complacency factor. Earlier in this book, it was mentioned that complacency occurs when after doing something well, due to much discipline and preparation, you see how much less you can do the next time and still achieve the same result. Complacency and familiarity are close cousins. According to an old saying, familiarity breeds contempt; however, it is actually complacency that breeds contempt.

In some situations, service professionals are at first disciplined with a task; then, as they grow more and more familiar with their work, they experience indifference, complacency, and finally contempt. It is a common cycle.

Pool lifeguards minimize complacency by moving to a different lifeguard stand every 30 minutes. This change of location gives the lifeguard fresh eyes to help stay alert and to spot danger in a pool.

On the job, complacency is a leading cause of accidents and injuries. Awareness and safety concerns are put on the back burner when service professionals become complacent in their jobs. The technician who works with high-voltage electricity must never become complacent about danger. The same rule applies to following up.

Most important, following up with customers shows that you care—and isn't that what customers really want? Customer service in its purest form is the act of one person helping another. Treat customers the way you wish to be treated.

Be Responsible

In my work with service companies, I have found that the "blame game" phenomenon exists. It is a result of field employees and inside employees who blame each other for bad outcomes. The best way to avert the blame

"Customer service in its purest form is the act of one person helping another. Treat customers the way you wish to be treated."

game is with a responsible attitude of taking ownership for each and every service event. This approach does not always come naturally. In most cases, it requires practice—similar to performing a role.

I urge service professionals to perform their role and thereby manufacture their own responsible attitude. How? By acting out the optimal desired behavior first rather than waiting and hoping for an improved mind-set and its associated behavior, they can artificially create a responsible attitude. It's called fake it until you make it.

A second skill set required for minimizing outcomes of the blame game is for a service professional to invest more time in listening and less time in talking, interrupting, and making excuses. Customers are much easier to handle when they get to talk, vent, and finish explaining their reason for calling. It is unfortunate, but in many service companies, customers are often interrupted in the interest of expediting a service call. My experience has taught me that interrupting customers results in just the opposite outcome. A customer who has been interrupted will return to the beginning of the complaint and start over again. Customers do this because they believe that the service professional is not listening. Like the responsible attitude, good listening skills must be rehearsed and applied. Service professionals who have been trained to be empathetic listeners achieve world-class service status.

Being responsible and taking ownership of problems is hard work. If it were easy, then everyone would be doing it.

Summary

Here are a few principles that service professionals can employ to combine fearlessness with finesse when working with customers:

- Busy service professionals engaged in multitasking activities will experience mind-drift numerous times within a day.

- The average person experiences mind-drift for about 30 percent of their waking day.

- Our ability to hear, qualify, and understand increases exponentially when we do three things: listen, take notes, and echo what was spoken.

- Sometimes it's best to turn off a portable communication device. Service professionals must discern when their listening skills need to be sharpest and then take the necessary precautions.

- Listening skills are diminished when a service professional's bias results in a preconceived notion about others.

- Service professionals must always practice restraint and not become defensive when these bias situations arise.

- In all customer encounters, service professionals invite the type of behavior that they convey toward customers.

- It's best to allow others to finish their complete thought and then ask a therapeutic question designed to open up a constructive dialogue. An example of a therapeutic question is, "Can you help me understand why you feel that way?"

- A handwritten "Thank you" card is an inexpensive, yet profound gesture that customers will remember.

- The law of reciprocation possesses awesome strength, often producing a yes response to a request that, except for an existing feeling of indebtedness, surely would have been refused.

- Upsell to an already receptive buyer to enhance the value of his or her purchase.

- Strategic follow-up, as part of a standard operating procedure, is a key differentiator among service professionals.

- Be responsible and take ownership of problems—it may not be your fault, but it is your job.

Appendix

Checklists and Forms

Communication Checklist

- ☑ Say "May I," Please," and "Thank You."

- ☑ Be prepared to convey value and benefits when customers focus on price.

- ☑ Minimize verbal junk—be clear, concise, and comprehensive.

- ☑ Refrain from editorializing—say less, listen more.

- ☑ Be prepared to contain adverse situations.

- ☑ Speak slowly when recording a voice-mail message.

Listening   Checklist

☑ **Hear, qualify, and understand what customers convey.**

☑ **Remember a customer's name.**

☑ **Clarify a customer's response when your mind drifts.**

☑ **Be objective and listen with little or no bias.**

☑ **Maximize empathy first and then convey expertise.**

☑ **Ask a customer if he or she has any questions, then listen carefully.**

☑ **Detach emotions and listen earnestly.**

Organization Checklist

☑ Make a pre-arrival phone call, and be proactive.

☑ Less clutter results in more mental clarity.

☑ Keep healthy foods in the truck for optimal attitude and performance.

☑ Be mindful of cognitive capture and its impact on mental awareness.

☑ Complete all paperwork, and do it right the first time.

☑ Resist the urge to procrastinate.

Cleanliness Checklist

☑ Be mindful of personal hygiene.

☑ Carry moistened hand wipes and breath mints in the truck.

☑ Keep an extra clean shirt in the truck.

☑ Wear shoe covers in a customer's home.

☑ Wear company-issued attire for the sake of congruency.

☑ Keep truck clean and orderly.

☑ Clean up before departing from a customer's property.

Attitude Checklist

☑ Make a positive first impression.

☑ Start the day as an optimist and keep that attitude all day.

☑ Smile and get into a positive frame of mind.

☑ Let the customer be right.

☑ Good eye contact and pleasant facial expressions speak volumes.

☑ Show appreciation for a customer's loyalty and trust.

☑ Resist the urge to cut corners and become complacent.

Company Dress Code Policy

Our company's goal in establishing a dress code is to allow our employees to work comfortably and safely. However, our employees must project a professional image to our customers, vendors, and the community at large.

Because not all attire, hairstyles, tattoos, jewelry, and piercings are suitable for the workplace, these guidelines will help you understand what is appropriate attire at work. Looking professional helps us convey a professional appearance to those whom we serve.

♦ **Shirts and blouses**

♦ **Pants and jackets**

♦ **Hairstyle and hygiene**

♦ **Makeup and fingernails**

♦ **Tattoos**

♦ **Jewelry and piercings**

♦ **Hats and head covering**

♦ **Shoes and footwear**

Write a commonsense expectation for each dress code item on the left.

Place a signed copy of this form in each employee's personnel folder.

Hold employees accountable.

Some service companies have varying degrees of tolerance, so establish criteria that are congruent with your company's culture and image.

I, the undersigned, agree to abide by the above policy, understanding the consequences for failure to comply, so that I may do my part to enhance the company image, my image and personal safety, and satisfy the customers.

Manager signature: _____ Date: _____

Employee signature: _____ Date: _____

Employee Expectation Guidelines

Our company's goal in establishing employee expectations is to ensure that customers are served and that our business can function smoothly and efficiently. When one employee fails to do his or her part, this disrupts the workflow and results in a disproportionate workload that compromises our standard operating procedures.

Employees are expected to work within the parameters of the following guidelines:

♦ **Be on time for work.**

♦ **Obey company policies.**

♦ **Respect coworkers.**

♦ **Be flexible during peak customer demand cycles.**

♦ **Keep your truck clean.**

♦ **Always tell the truth.**

♦ **Obey traffic laws and be a courteous driver.**

♦ **Attend in-house training classes and meetings.**

Write a commonsense expectation for each item on the left.

Place a signed copy of this form in each employee's personnel folder.

Hold employees accountable.

Managers should establish a performance review schedule and stick to it. Employees will respect what is expected from them.

I, the undersigned, agree to abide by the above guidelines, understanding the consequences for failure to comply, so that I may do my part to be a productive employee who will contribute to our company's success and exceed our customer's expectations.

Manager signature: _____ Date: _____

Employee signature: _____ Date: _____

Technical Service Manager Guidelines

Our company's goal in establishing Technical Service Manager Guidelines is to ensure that our systems and processes serve our customers and employees so that our company thrives and is differentiated in the marketplace. The role of management is to build and maintain a stable infrastructure that enables our employees to become more successful. Technical managers must get things done through their subordinates, and this can occur only if managers delegate effectively. Therefore, ongoing training and assessment of employee skills is an essential management requirement.

continual and ongoing training must rotate among three business topics: (1) procedures and processes; (2) technical skills and certifications, and (3) soft skills and problem solving.

Technical managers are expected to work within the parameters of the following guidelines:

♦ **Develop a stable infrastructure (processes and systems).**

♦ **Provide continual and ongoing training classes.**

♦ **Delegate effectively.**

♦ **Be accessible to subordinates.**

♦ **Lead by example.**

♦ **Use customer feedback to improve systems.**

Write a commonsense expectation for each item on the left.

Place a signed copy of this form in each technical manager's personnel folder.

Hold technical manager accountable.

Company owner should establish a performance review schedule and stick to it.

I, the undersigned, agree to abide by the above guidelines, understanding the consequences for failure to comply, so that I may do my part to be a productive employee who will contribute to our company's success and exceed our customer's expectations.

Company owner signature: _____ Date: _____

Technical manager signature: _____ Date: _____

Index

DATE DUE
